Origin and Evolution of the

The Royal Society of Canada
The Canadian Institute for Advanced Research

Origin and Evolution of the Universe: Evidence for Design?

Introduction by
Alan H. Batten, F.R.S.C.

Edited by
John M. Robson, F.R.S.C.

McGILL-QUEEN'S UNIVERSITY PRESS

Kingston and Montreal

ISBN 0-7735-0617-9 (cloth)
ISBN 0-7735-0618-7 (paper)
Legal deposit fourth quarter 1987
Bibliothèque nationale du Québec
Printed in Canada

The papers in this volume
originated in a Symposium at McGill University,
30 May–1 June 1985,
sponsored by the Royal Society of Canada
with the cooperation of
the Canadian Institute for Advanced Research.

This book has been published with the help of a grant from the
Canadian Federation for the Humanities, using funds provided by
the Social Sciences and Humanities Research Council of Canada.
Additional support was received from the Canadian Institute
for Advanced Research, one of the sponsors of the Symposium.

Canadian Cataloguing in Publication Data

Main entry under title:
Origin and evolution of the universe
ISBN 0-7735-0617-9 (bound). – ISBN 0-7735-0618-7 (pbk.)

1. Cosmology – Congresses.
2. Cosmogony – Congresses.
3. Life – Origin – Congresses.
4. Teleology – Congresses.
I. Robson, John M., 1927–   .
II. Royal Society of Canada.
III. Canadian Institute for Advanced Research.

QB980.O74 1987   523.1'2   C87-094104-6

This volume is dedicated to the memory of

EDWARD R. LLEWELLYN-THOMAS, F.R.S.C.,

1917–84

Engineer, Physician, and Poet

Born in England, "Tommy" Llewellyn-Thomas early went to sea in the merchant service and qualified as an electrical engineer in 1938. During the Second World War he served on merchant ships and later in the Royal Signal Corps and the Royal Electrical and Mechanical Engineers. From 1946 to 1951 he was in Malaya setting up a telecommunications link between Australia and London. In 1951 he entered McGill University as a medical student and graduated in 1955. After a period in general practice in Nova Scotia, he held a variety of positions and eventually became Associate Dean of the Faculty of Medicine in the University of Toronto. He was elected a Fellow of the Royal Society of Arts in 1970, and of the Royal Society of Canada in 1975.

He was keenly interested in this symposium and helped to generate support for it in the Royal Society of Canada. His death, less than a year before the symposium was held, deprived us of one who would have been amongst its most enthusiastic participants.

Contents

ALAN H. BATTEN

Introduction

The idea of this symposium came to me as a result of reading Paul Davies' recent book, *The Accidental Universe* (1982), in which he explains some of the insights that have emerged from bringing together astronomical cosmology and quantum physics in the study of what might have happened at the very earliest beginnings of the universe as we know it. If we may assume, as most astronomers and physicists now do, that the expansion of the universe, which we infer from our modern observations of other galaxies, began in a single event somewhere between 10 and 20 thousand million years ago, then the circumstances of that event seem (at least in retrospect) to have been very finely "tuned." If any of several physical parameters had been only slightly different, no form of intelligent life, recognizable as such by us, could have come into existence. For example, a relatively small difference, either way, in the ratio of gravitational to nuclear forces would have made the formation of stars impossible. As far as we know, the interiors of stars are the only places where the heavy elements (especially carbon) that are essential for our bodies may be made. Further (again, as far as we know) it is only on an aggregation of rock and iron, placed at the right distance from a relatively cool small star, that the conditions needed for any form of life are found. For both these reasons, therefore, a universe without stars would be a universe without life.

Davies' book, however, was not the only stimulus. A year or two earlier, I had read Lovelock's *Gaia: A New Look at Life on Earth* (1979) and, at about the same time, A.H. Cook's presidential address to the Royal Astronomical Society. In "Geophysics and the Human Condition" (printed in the *Quarterly Journal of the Royal Astronomical Society*

20 [1979]: 229-40), he explained how many of the features of our physical life are dependent on the kind of surface that we find on Earth. It seemed to me that both these authors, each from a different branch of science, were coming close to presenting some form of teleological argument, without explicitly advancing it. Perhaps they had their reasons for stopping short (I did not know until after the proposal for this symposium had been made that Davies planned a sequel to his book), but I thought that it would be useful to bring together people from a wide spectrum of disciplines who were working along these lines. If our discussions were indeed to touch upon the teleological issue, it was obvious that we should include theologians and philosophers, for they have studied the matter for centuries, or even for millennia. Plato knew of such ideas, for he mentioned them in the tenth book of *The Laws*, although he was apparently far from convinced by them. We might see even older elliptical and poetic expressions of a similar idea in the famous opening verse of the nineteenth Psalm, and in the Pythagorean doctrine of the music of the spheres. Indeed, the very word *cosmos* has a root meaning of "order" or "arrangement," and we could argue that ever since human beings have been able to express their reflections on the universe, they have seen in it some evidence for design.

It is, of course, obvious that if conditions in the early universe had not been right for the ultimate emergence of life, we would not be here to ask questions about it. This is the difficulty that the teleological argument must always face; in my lifetime it has been widely held (and at times I have held too) that Darwin's theory of natural selection has dealt teleology a fatal blow. If only those creatures survive that are best fitted to do so, then inevitably – even by chance – they will appear to have been designed to fit their environment. This objection is not original with Darwin; it is as old as teleological ideas themselves, and is found explicitly in the surviving writings of the pre-Socratic philosopher Empedocles. Another way of discounting the appearance of design has been to postulate the existence of "many universes" of which we, not surprisingly, observe the one in which we are able to exist. "Many universes" is perhaps an unfortunate phrase, but it will have to serve until a better one is found. But even if, as Hume suggested, "Many worlds might have been botched and bungled, throughout an eternity, ere this system was struck out," there is still a possibility that all were tried in order to bring into being one in which we could wonder "Why?"

If the argument from design, and the objections to it, are so venerable, is there any point in our meeting to discuss them near the end of the twentieth century? I believe there are two differences between modern forms of the argument and older ones. In the English-speaking world, at least, the eighteenth-century form has become classical. It had two exponents – Paley, who used it to persuade his readers of the validity of the argument, and Hume, who wished to refute it but who also gave a powerful exposition of it. Each chose a machine as the type of design. Since their mature lives were spent in Britain between the death of Newton and the beginning of the industrial revolution, this choice of analogy is hardly surprising, but it does not have much appeal now. The design we see now, if we do at all, is more like the "harmonies" that induced Kepler and Galileo to abandon the Ptolemaic system for the Copernican, or, indeed, like the music of the spheres; or it may be argued by some that design in the universe is more correctly compared to what we see in a living organism (another Humean idea). The other difference is more important. The Copernican harmonies could be appreciated by the learned, but they were not necessary to existence. The harmonies that we see now, whether in the universe as a whole, or in the fundamental particles of matter, are vital to us. Without them we would not be here; yet thousands of millions of years before we *were* here, they existed. This, I believe, adds an entirely new dimension to – if I may re-translate from the French version of our title – the question of intentionality.

It has become a truism that science is concerned only with efficient, and not with final, causes. Within its own terms of reference, science has certainly made great advances by ignoring the final cause; indeed, Galileo deliberately tried this approach. Nevertheless, Galileo, Kepler, and Newton, in spite of their conflicts with the reigning orthodoxies, were religious men living in a Christian Europe who surely believed that there were such things as final causes. It was only later that scientists seemed to forget that the chief reason they could find no evidence of a final cause was that they had stopped looking for it, having defined it to be outside the realm of science. When Napoleon questioned Laplace about the role of God in the creation of the Solar System, he is supposed to have answered: "Je n'ai pas besoin de cette hypothèse-là." I do not know whether the story is true, but it does illuminate the truth. It was after all in the early nineteenth century that materialism and atheism became fashionable amongst some scientists. Since then, however, we have seen the growth of the theory of evolution and of physical cosmology, and both are syntheses of such breadth and complexity that one doubts if the Aristotelian distinctions between different causes are applicable any longer to them. The "why?" of each of these two subjects is so inextricably bound with the respective "how?" that it must be almost impossible to study either and to remain neutral about the existence of a final cause. Scientists are being forced, by the very success of their decision to ignore final causes, to confront again the possibility that such causes exist. At least, if natural science is to be worthy once again of its old name of natural philosophy, we must face this question.

If we are persuaded by evidence for design in the physical universe, it necessarily follows that we are a part of that design. This places a responsibility on us. How great that responsibility is depends on our estimate of the possibility and frequency of life in the universe, and several of the papers in this volume have a bearing on the question. Until a few years ago, many astronomers displayed an almost facile optimism about it (I assume it *is* "optimistic" to suppose that there are many life-bearing planets in the universe); biologists, however, have always been more conservative. Recently, the pendulum seems to have swung to a more conservative position amongst astronomers as well; many have come to believe that we are the sole representatives of intelligent life in the universe. Dare we return to such a degree of anthropocentricity as to assert that there is an intention in the universe and that that intention was to produce intelligent beings like us? If so, it places an awesome responsibility on us all. We simply may *not* blow ourselves to bits. If any subject demands the combined efforts of all three of the great branches of philosophy – mental, moral, and natural – it is the theme of our symposium.

When all is said and done, however, no evidence for design, nor, indeed, acceptance of any of the traditional arguments for the existence of a deity, will bring us to a belief in the Christian God, or, for that matter, the God of Abraham, of Isaac, and of Jacob. There may be some here who regret that limitation on our discussion; others would not have joined us had they thought that this symposium was to be an occasion for preaching in disguise. The business of the Royal Society of Canada, after all, is not evangelism, but scholarship. Once again, it was Hume who pointed out that this kind of reasoning would lead only to natural religion, not to revealed religion. St Thomas Aquinas would have agreed with him completely on this, although I suppose the two men would have disagreed about almost everything else. The categories of Christianity, or of Judaism and Christianity, are so deeply entrenched in our culture that, when we begin to think about this kind of question, even those who consciously have rejected those categories, often use them unconsciously. A Buddhist, Hindu, or Moslem can perhaps jolt us into a different point of view, and for that reason we welcome a speaker with his roots in the Hindu tradition. Perhaps, among other things, this contribution will serve to remind us that, just as "cosmos" is a Greek word, so the notion of design in the universe is a particular part of the Greek and Jewish heritage of Christian Europe. Perhaps the question in our title is a consequence of the way in which we think these things. Yet of those moments when we ponder these ideas, we might say something Eddington said in a not very different context:

These were not moments when we fell below ourselves. We do not look back on them and say, "It was disgraceful for a man with six sober senses and a scientific understanding to let himself be deluded in that way ..." It is good that there should be such moments for us. Life would be stunted and narrow if we could feel no significance in the world around us beyond that which can be weighed and measured with the tools of the physicist or described by the metrical symbols of the mathematician. (*The Nature of the Physical World* [1928])

All symposia require the efforts of many people if their success is to be assured. As Chairman of the Organizing Committee, I would like to thank all my colleagues on that Committee, but especially Professor A. McKinnon, who worked hard on the local arrangements, and the officers of the Royal Society of Canada, especially Professor K.J. Laidler (Chairman of the Society's Committee on Symposia) and Mr E.H.P. Garneau (Executive Secretary) and his staff. The decision of the Canadian Institute for Advanced Research to join in sponsoring the symposium proved to be an important element in the realization of our aims, and I am grateful to the President, Dr J.F. Mustard, for his enthusiastic support. Financial support was also received from the Natural Sciences and Engineering Research Council and the Social Sciences and Humanities Research Council. To each of these bodies we express our thanks. Finally, we gratefully acknowledge the hospitality extended to us by McGill University, and express our thanks to the Principal, Dr David Johnston, and to all those who worked behind the scenes to ensure the smooth working of the physical arrangements for the meeting. We are greatly indebted to Rea Wilmshurst for preparing the final manuscript.

Origin and Evolution of the Universe

ROBERT H. HAYNES

The "Purpose" of Chance in Light of the Physical Basis of Evolution

Apologia

In my youth I was taught to be a devout Christian, but I soon became enamoured of the ethics and physics of Epicurus [1]. This transformation occurred as I was attracted by the explanatory and predictive powers of science. Theistic accounts of the origin, nature, and future of the world seemed, in comparison, to be inadequate, unsound, and, to a naive realist, contrary to the plain evidence of one's senses.

I remain entranced by the Epicurean view of the universe unfolding through the indeterminate swerve and collision of "atoms" in the void. Chance events in the atomic and molecular domain lie at the root of the structure, organization, and diversity that we see around us. In the world of living organisms the manifest "purpose" of chance is to create the genetic information that makes possible adaptation and design, and to confer uniqueness upon individuals. In a completely deterministic universe, as Newton rightly observed, "blind metaphysical necessity, which is certainly the same always and everywhere, could produce no variety of things" [2].

The universe and its parts, both physical and biological, have been compared often with machines. Clocks, heat engines, and computers variously have been invoked as models for essential features of the world. To me at least, a more apt metaphor is provided by a game of chance, played out on a vast, microcosmic wheel of fortune. The game, of course, has rules [3]. These are the laws of nature through which the consequences of chance atomic events are ordered, statistically and dynamically, into coherent, but transient, structures, as diverse as crystals, organisms and stars. Such structures may be regarded as players in the game. Atoms and molecules, particles and fields, constitute the basic hardware from which both wheel and players are fashioned; usable energy allows the game to proceed. We sentient beings are the only players known to us who also are observers. Since entering the game we have developed, through science, a remarkable capacity to discern its rules. However, with self-consciousness there has come also a deep psychic need to impute some transcendent meaning and purpose to the world. Unfortunately, it is irrational to ask *why* the universe exists because it is impossible logically to obtain an answer: any purported solution, theological or otherwise, simply invites, in infinite regress, a further question "why?" Chronic addiction to such unanswerable questions can rend the fabric of the mind.

Not only the chance events of quantum mechanics, but also the particular mathematical forms taken by the deepest laws of physics appear to be inexplicable (or tautological, insofar as they are based on mathematical identities). If any ultimate *raison d'être* for the game exists, it is beyond the grasp of finite minds. The only evident "purpose," for biological players at least, is to develop evolutionary strategies that allow them to "win," and thereby remain in the game as long as possible. In the last analysis, however, the appearance and disappearance of individuals, species, and indeed the game itself, seem to be governed by a turning wheel of fortune.

Argumentum

Enunciation of the mechanism of evolution by natural selection [4] undermined the ancient notion that evidences for design in nature, and the adaptation of biological form to function, necessarily imply the existence of a transcendent Designer. However, the sustaining physical sources of evolutionary change lie not in selection but rather in the occurrence of variations among organisms, and the greater the range of genetic variability, the greater is the scope for natural selection to act. Indeed, Darwin himself emphasized that in the absence of heritable variation, selection can do nothing [5].

Heredity is a manifestation of the *stability* of genes and chromosomes from one generation of cells to the next. Variation is a manifestation of the *instability* of genes (mutation) and of their recombination, rearrangement, and reduplication within the genome. The molecular basis of genetic stability and change, and the intimate relation between these rival processes, is rooted in the chemical nature of the genetic material itself, and the biochemical mechanisms for its replication, repair, and recombination.

The formation of normal progeny, and the integrity and longevity of species, depend on the astounding accuracy with which the genetic material (DNA, or deoxyribonucleic acid) can be replicated and preserved, from generation to generation. This great stability results from the efficient operation of a complex system of enzymatic mechanisms that protect DNA against damage, that repair it when damaged, and that promote fidelity during its replication [6]. The genetic machinery of the cell is a beautiful example of a highly reliable, dynamic system built from vulnerable and unreliable parts.

Despite the sophistication of the devices that promote genetic stability, they cannot operate perfectly at all times and under all circumstances. Mutations, arising ultimately from many diverse sources of uncompensated molecular "noise," must, and do, occur at some low frequency. However, because of the great precision of replication, those mutations whose effects are not inconsistent with cellular metabolism, organismal development, function and behaviour, will be replicated accurately and recombined in various patterns along with pre-existing genes. Ultimately, they may become established in the gene pool of the species. Through largely stochastic processes of evolutionary "tinkering," [7] natural selection orders and preserves the adaptive results of this "bootstrapping" interplay of random mutation, recombination, and accurate replication of the genetic material. Thus, in the realm of life, chance and necessity are intimately coupled in the molecular basis of genetic stability and change.

Quantum mechanics [8], thermodynamics [9], and molecular biology [10] collectively reveal that the laws governing the fundamental transformations of matter and life are, at bottom, statistical in character. They are the laws of aggregates and averages, based upon chance events, statistical fluctuations, and molecular accidents. Contemporary physicists seriously consider that the initial singularity of "big bang" cosmology may have appeared spontaneously "from nowhere" as a quantum fluctuation of the vacuum (in the sense of quantum field theory) [11]. It would indeed appear that, in the words of Milton, "chance governs all." Yet from this primeval atomic ferment biological order and sentient beings have emerged. We have good reason to be awed by the wonder and numbed by the absurdity of it all. In the narrative that follows I give a brief account of the historical development of these ideas in the biological domain. This leads to some unrefined speculation on the possible adaptive significance of belief systems in man. I conclude with a short homily on the perils of simplicity, and the virtue of humility, in the paradoxical world of mind and matter in which we find ourselves [12].

Narratio

ADAPTATION: NATURAL THEOLOGY AND NATURAL SELECTION

In 1858, Charles Darwin and Alfred Russel Wallace provided the first, and to this day the only observationally established, naturalistic explanation for the teleonomic adaptation of biological form to function [4]. Such adaptations bear every hallmark of exquisite design. For centuries, philosophers, artists, mystics, and sensitive people in all walks of life have been variously awed, inspired, and intrigued by the beauty and mystery of the living world. How could one doubt that the deliberate, purposeful character of so much daily human activity was but a reduced reflection of some larger purpose in Nature, if not the will of a predestining God? Indeed, many of the most profound and romantic expressions of the human spirit have come from the poetic contemplation of biological phenomena.

Prior to the enunciation of the mechanism of natural selection, scientists and theologians alike saw God not only through the eye of faith, but also in the light of nature. For example, in 1691, John Ray, the father of British natural history, published his celebrated book, *The Wisdom of God Manifested in the Works of the Creation*. Ray described the "multitudinous variety and perfection" of living organisms, the "beauty and minuteness" of their parts, and gave many now classic examples of anatomical and physiological adaptations in plants and animals. Furthermore, he argued that it was futile to attribute the formation of the world to chance events.

In 1802, William Paley published his famous work, *Natural Theology, or Evidences of the Existence and Attributes of the Deity Collected from the Appearances of Nature*. Paley expressed the hope that his effort would provide "strong foundation for the belief that God Almighty wills and wishes the happiness of his creatures," but cautioned that "adequate motive must be supplied to virtue by a system of future rewards and punishment."

It is easy today to mock the views of Paley and, even more, those of the later authors of the Bridgewater Treatises. However, we should not forget that Boyle, Hooke, and Newton, all towering figures of the scientific revolution, also were ardent votaries of natural theology and the argument from design.

The theory of natural selection provided biologists with a plausible mechanism for the origin and evolution of species. In its original form the argument proceeds in stages as follows: first, Darwin observed that organisms tend to multiply geometrically, even in the face of limited resources for survival. He then asked how the resulting Malthusian struggle for existence will act with regard to the variations that are observed to occur among the individuals of all species. His answer, in his own words, was:

If such [variations] do occur, can we doubt ... that individuals having any advantage, however slight, over others, would have the best chance of surviving and of procreating their kind? On the other hand, we may feel sure that any variation in the least degree injurious would be rigidly destroyed. This preservation of favourable variations and the rejection of injurious variations, I call Natural Selection. Variations neither useful nor injurious would not be affected by natural selection, and would be left a fluctuating element, as perhaps we see in the species called polymorphic. [5]

Through the lens of the "retrospectoscope," an instrument almost as powerful as the eye of faith, Darwin's statement of his famous principle is truly remarkable. It resonates with the major disputes that have arisen during the subsequent development of evolutionary theory. Its most significant feature is that it combines physical causes, that is, reproduction, heredity, and variation, with biological consequences, that is, competition and selection. Thus, evolution can be viewed as a two-step process that proceeds from the genesis of heritable variation among individuals to selection at the population level.

As Darwin correctly saw, natural selection is the immediate or proximal cause of evolutionary change. However, selection is not a force of nature analogous to gravity or electromagnetism. Rather it arises as a consequence of the pre-existence of hereditary variation and reproductive excess within populations. Heritable variations occur randomly in each generation and are not passed on with equal probability from one generation to another. Through the action of the opportunistic process of natural selection, unfavourable variations are lost and favourable variations proliferate. Species gradually become well adapted to their physical and biological environments and, through many selectively mediated adjustments, to one another within the global ecosystem.

At the genetic level selection should not be regarded as some mere sieve of "bean-bag" genetics. In response to environmental challenges, variant genes are ordered and integrated through selection into concordant groups and patterns. This process produces the adaptive relations of form to function that we see in nature. Thus, organisms as they exist today reflect the moulding of accumulated heritable variation by natural selection, but without variation there would have been nothing to mould.

In Darwin's day, the basic mechanisms of heredity and variation were badly misconstrued. It was thought that heredity entailed a "blending" of the germinal fluids of the parents in sexually reproducing species. On this basis, heredity was considered to be a conservative process that suppressed variations within one or a few generations of their appearance. Fleeming Jenkin, Darwin's most serious and effective critic, argued that variations could not be preserved by natural selection against the "swamping effect" of the presumed blending process [13]. Indeed, in 1879, the British writer Samuel Butler could say, with considerable justice, that "in not showing us how the individual differences first occur, Mr. Darwin is really leaving us absolutely in the dark as to the cause of all modification – giving us an 'Origin of Species' with the 'Origin' cut out." Darwin realized that this was the greatest single deficiency in his theory and he went to great pains to overcome it, particularly in the elaboration, in 1868, of his highly speculative "provisional hypothesis of pangenesis" [15].

Darwin argued that heritable variations occur at random with respect to any adaptive significance they might have for the species in question. On the other hand, he was unequivocal in stating his belief that variations are not caused by chance or accidental events. He attributed their appearance of spontaneity to the then current ignorance of the underlying laws of variation that he assumed must exist, and someday would be discovered [5, 15].

As Darwin brought the *Origin of Species* through its six editions, he increasingly invoked, as a source of variation, the effects of "use and disuse," as entailed in the soon to be discredited Lamarckian notion of the inheritance of acquired characteristics. Indeed, it has since become clear that no causal relation, or coupling, exists between the molecular mechanisms of mutagenesis and processes of selection at the phenotypic level. The *particular* genetic variants that arise in any one generation do not constitute a response to environmental exigencies, nor to the advance of any kind of orthogenetic evolutionary "progress" [16].

Despite the problem posed by the unknown origin of variations, the fact that biological adaptations, with all their complexity, could be readily explained by a naturalistic mechanism had an immediate and far-reaching impact on the *Weltanschauung* of simple and sophisticate alike. Even today Darwin's ideas seem as troublesome to the pious as they were in 1859. Yet that great zoologist and staunch supporter of Darwin, Sir Ray Lankester, saw that in a curious way, teleology was "refounded, reformed and rehabilitated" in the very concept of natural selection itself. "Every variety of form and colour was urgently and absolutely called upon to produce its title to existence either as an active useful agent, or as a survival of such usefulness in the past" [17]. Again, to paraphase D'Arcy Thompson, writing in 1917, one reaches a "teleology without a *telos*," a "final cause" based on a sifting of better from worse, and a world that, willy-nilly, is tending to become the best of all possible worlds [18]. Despite the superficiality of this "Panglossian paradigm," to borrow the phrase of Stephen Gould and Richard Lewontin [19], much traditional evolutionary biology has been informed largely by the naive pan-selectionism of the adaptationist program. However, the rise of quantum mechanics and molecular biology has made it clear that the deep philosophical implications of evolution must be sought, not only in the processes of adaptation through selection, but, more importantly, through an understanding of the physical basis of evolution as embodied in the molecular mechanisms of heredity with variation.

THE CHROMOSOMAL BASIS OF HEREDITY AND VARIATION

Two years prior to the publication of the *Origin of Species*, an Augustinian monk, Gregor Mendel, began to analyse the quantitative distribution of hereditary traits among the offspring of hybrid pea plants grown in his monastery garden. On the basis of simple numerical ratios characterizing the frequencies of reappearance of parental traits in subsequent generations, he concluded, in 1865, that there must reside in plants particulate units, or "genetic factors," later christened "genes," that control the hereditary attributes of organisms. Furthermore, he deduced that there exist two such genes corresponding to each of the traits he studied, one of paternal, the other of maternal, origin. He showed that these two homologous units do not blend during their residence in any one individual. Rather, they segregate intact, one from the other, before being passed on through sperm or egg to the next generation [20]. Unfortunately, Darwin never

heard of Mendel's atomistic theory of inheritance. It remained hidden
on dusty library shelves until its dramatic rediscovery by Hugo de Vries
and others at the turn of the century. The chromosomal basis of
heredity and variation was worked out during the subsequent
development of classical Mendelian genetics that was led by William
Bateson [21] in England and Thomas Hunt Morgan in the United States
[22].

Mendel had no idea of the location or chemical nature of his genetic
factors. However, in 1902, the American W.S. Sutton deduced, by
combining genetic with cytological data, that genes must abide in the
thread-like chromosomes that had been found in the cell nuclei of
sexually reproducing species. Through genetic studies of the fruit fly
Drosophila melanogaster, Morgan and his colleagues showed that genes
are linked in linear array along these chromosomes. Indeed, they were
able to develop genetic "linkage maps" that revealed the relative
positions of particular genes on particular chromosomes. The classical
geneticists also showed that differences between individuals could
arise in two ways. First, by gene and chromosomal mutation; and
second, by the random assortment and recombination of maternal and
paternal chromosomes during the "reduction division" (meiosis) of
the two homologous chromosome complements that are involved in
sperm and egg formation. Gene, or "point" mutations arise from
changes in the detailed chemical structure of individual genes.
Chromosomal mutations arise from changes in the number of genes
in chromosomes (deletion, duplication), in the location of genes within
chromosomes (inversion, translocation), and in the actual number
of chromosomes per cell.

Point mutations are of special interest because they constitute rare
exceptions to the normal process of highly accurate gene replication
from one cell generation to the next. The alternate forms of any given
gene that arise in this way are called "alleles." Any two of several
different alleles may constitute an homologous gene pair. If the two
alleles are identical, the pair is said to be "homozygous," if different,
"heterozygous." It is the existence of heterozygosity that gives genetic
significance to recombination and the random assortment of homologous chromosomes in meiosis. Because the number of genes even in
simple organisms is measured in the thousands, a small amount of
heterozygosity is sufficient to generate an astronomical number of
potentially variant gametes. For example, if an organism with 10,000
gene pairs is heterozygous for only one per cent of them, the number of

genetically different gametes it could, in principle, produce is 10^{30}! Clearly, every individual formed from the union of sperm and egg, drawn randomly from parents exhibiting normal levels of recombination, will be unique in the entire history of the species. Indeed, in recent years it has been shown that the amount of heterozygosity in organisms, and the number of alternate alleles for many genes, is far greater than ever imagined prior to the development of the sensitive molecular techniques needed for their detection. Thus, there exists an enormous reservoir of genetic variability within the gene pools of species upon which selection can act in evolution. This accounts for the historical ability of breeders, using artificial selection, to produce, albeit within limits, remarkable changes in the characteristics of domestic plants and animals over relatively few generations.

NEUTRAL ALLELES AND THE SYNTHETIC THEORY OF EVOLUTION

Soon after the rediscovery of Mendel's work and the observation that mutations having dramatic effects on organisms could occur, Darwin's exclusively gradualistic view of evolution by the slow accumulation of favourable variations through selection fell somewhat into eclipse. A myopic, but bitter, debate ensued in which Mendelians were pitted against Darwinians over the relative importance for evolution of marked, discontinuous variations, and even "hopeful monsters," as opposed to selection acting on small, continuous variations. The situation was largely resolved by the early 1930s with the development of mathematical population genetics by S.S. Chetverikov [23], R.A. Fisher [24], J.B.S. Haldane [25], and Sewall Wright [26], and the subsequent promulgation of the neo-Darwinian "synthetic" theory of evolution by Theodosius Dobzhansky [27] and others [28]. The population geneticists showed that Mendelian genetics and Darwinism are compatible with one another. Through their work, evolution became identified at the genetic level with changes in allele frequencies in the gene pools of populations under the influence of mutation, migration, selection, and random genetic "drift." The latter, purely statistical, phenomenon arises from the vagaries of intergenerational sampling of gametes for fertilization in finite populations. In the absence of such influences, gene frequencies would remain constant from one generation to the next, and evolution would not occur, except by new mutations.

Considerable controversy has developed over the relative importance of these four factors in determining the genetic structure of
populations and in promoting changes in gene frequencies [29]. The
more widely accepted view, developed primarily by Dobzhansky, has
been that selection is of overwhelming importance. Genetic variability,
in particular, heterozygosity in sexual species, generally has been
considered to be rigorously maintained by various forms of positive
"balancing selection" acting at the genotypic level [30]. However, the
discovery of a surprisingly large amount of allelic variation in populations has led to the elaboration of a more stochastic view of genetic
variability known as the "neutral theory" of molecular evolution [31].

The neutral theory asserts that almost all variation is selectively
neutral, or nearly neutral, with respect to existing genes, and is
maintained primarily by mutational input and random drift [31, 32].
New mutant alleles are, for the most part, functionally synonomous
with existing genes. It is believed that such mutants arise frequently
and that they "percolate" for varying periods of time through the gene
pool with fluctuating frequencies. A few eventually become "fixed,"
that is, they displace, by random drift, other pre-existing alleles; but
the majority are lost within a few generations of their appearance, also
through the stochastic effects of drift. Natural selection generally acts
negatively to reduce variability by eliminating phenotypic extremes.
The role of positive selection in determining the course of adaptive
evolution is by no means denied by the neutral theory, and initially
neutral alleles can change in selective value as environments change.
The theory was designed primarily to explain the existence of so much
variability at the molecular level, though it also lends further emphasis
to the mechanistic decoupling of variation and selection. Nonetheless,
the internecine debate between selectionists and neutralists has been
sharp, though it would appear that these competing views can be
accommodated within a more relaxed version of the synthetic theory
[33]. Despite the controversy over neutralism, it remains agreed that
whatever the relative roles of selection, mutation, and drift in
determining the genetic "fine structure" of populations, random
mutation and recombination are the ultimate sources of variation in
evolution.

The large standing variation in populations, from which most
deleterious alleles have already been removed by selection, rather than
new mutations, provides the immediate material for phenotypic
adjustment to changing circumstances [34]. Even though selection acts,

in most situations, to reduce (extreme) variability, a small mutational input, together with recombination, is sufficient to provide an enormous amount of potential variation.

It must be emphasized that new mutations which are not neutral, but have significant effects, are far more likely to be deleterious than beneficial to their hosts, simply because some degree of adaptation would have been achieved historically by the species in question. The working of a well-tuned machine is unlikely to be improved by letting a child play at it with screwdriver and spanner. Thus, sexually reproducing organisms do not increase their mutation rates as an adaptive strategy. Such a manoeuvre, in all probability, would weaken if not extinguish the species [34].

NEW GENES AND GENOMES

Three main categories of genetic change are involved in evolution, all of which involve stochastic processes at various levels of biological organization. They are: first, changes in the relative frequencies of existing alleles in populations; second, generation of new alleles and the loss of others; and third, creation of new genes and new patterns of gene expression. To the extent that evolution depends on the first of these processes, its rate is largely independent of the mutation rate and is controlled primarily by selection. To the extent that it depends on the generation of neutral alleles, or new favourable mutations, the rate of evolution is proportional to the mutation rate [34]. However, the acquisition of new genes and new functions is the most prominent feature of evolutionary divergence and diversification above the species level.

The single chromosome of the bacterium *Escherichia coli* contains almost four thousand genes, of which about one thousand have been identified thus far. The human genome, which comprises forty-six chromosomes of unequal length, is thought to contain something between 20 and 100 thousand genes. A variety of mechanisms are involved in genomic expansion. These include the fusion and duplication of pre-existing genes [35] and parts of genes [36], possibly brought about by various modes of non-reciprocal recombination and replicative transposition of the genetic material within genomes [37], as well as (more rarely) the "horizontal" transfer of genes among species [38]. The redundancy of duplicate genes could allow one of them, being free from the constraints of selection, to diverge in structure and take on novel functions [35].

Not only does the number of genes vary widely throughout the plant and animal kingdoms, but the variation in the total length of the genomic DNA is even more marked: the genomes of certain plants and fish are one hundred thousand times longer than those of some micro-organisms. In higher plants and animals, as little as ten percent (or even less) of the genome may be devoted to "classical" genes that code for proteins, even allowing for the existence of non-coding "control" regions in chromosomes. It has become apparent that the genome is a remarkably "plastic" structure and that a variety of mechanisms exist whereby the genetic material can be rearranged and altered in composition and length. Obviously, genetic engineering has been going on routinely in nature long before molecular biologists even contemplated such things! Confronted with so many ways in which genomes can change in evolution, the main problem becomes that of determining how and why they actually have changed in various phylogenetic lines of descent.

We have come, therefore, to a picture of the production of genetic variation by a vast, probabilistic "reshuffling," at each generation of the many allelic forms of genes that have accumulated slowly, by random mutation, in the gene pools of species over evolutionary time. The recombinant gene patterns that emerge from this genetic lottery can vary in overall selective value from one individual to another. However, even though spontaneous mutations are the basic source of variation, mutation rates must be kept extremely low since most new mutations are deleterious rather than beneficial. Heritable variation, as understood within the framework of Mendelian genetics, reveals two antithetical aspects of the behaviour of genes. Heredity, and genomic expansion in evolution, depend upon a capacity for the precise replication of genes. On the other hand, variation depends on mutation or the instability of genes. This paradoxical requirement for both genetic stability and change points up a fundamental dialectical issue in molecular biology. It has been resolved over the past thirty years through the discovery of the chemical nature of the gene, and the various enzymatic mechanisms that effect the replication, repair, and recombination of the genetic material. These remarkable advances in molecular biology complement the revolution begun by Darwin. Furthermore, they have put in our hands, through the development of recombinant DNA technology, the possibility not only to understand our genetic past, but also to manipulate our genetic future [39].

MOLECULAR MECHANISMS IN GENETIC STABILITY
AND CHANGE

1 *Physical Theory of Gene Stability*

In 1909, the Danish geneticist W. Johannsen introduced the term "gene" to science and showed that mutations are rare, sudden and discrete events that cause genes to pass from one stable state to another. In 1919, H.J. Muller, the American geneticist, who later won the Nobel Prize for the first artificial transmutation of genes, showed that the temperature coefficient for spontaneous mutation in *Drosophila* was much larger than that observed for most common physiological processes [40]. Early searches for chemical mutagens gave negative or equivocal results, and by the mid-thirties the only known mutagenic agents were the high energy quanta of X-rays and ultraviolet light [41]. The extreme rarity of spontaneous mutation, together with the unusually high energy threshold for the process, made the gene appear to be a very strange bird indeed. All of this led Max Delbrück, one of the great physicist-fathers of molecular biology, to propose in 1935 that gene stability and the high temperature coefficient for mutagenesis could be explained on the basis of the Polanyi-Wigner theory of molecular fluctuations [42]. Thus Delbrück suggested that genes might be composed of offbeat, extraordinary molecules, perhaps constituting some special, hitherto unrecognized, state of matter. Mutations would be chance quantum events, rendered rare because of the high energy thresholds for the appropriate isomeric transitions.

This theory had much appeal for the Austrian physicist, Erwin Schrödinger, the father of wave mechanics and the author of the famous equation of quantum theory that bears his name. He was puzzled by the continued existence of a peculiar deformity, known as the "Habsburg lip" in succeeding generations of rulers of the Austro-Hungarian Empire. How could the mutant gene that gives rise to this imperfection have remained for centuries in the royal gonads unperturbed by the disordering tendencies of heat motion? Delbrück's theory provided a respectable physical answer. Thus, in his famous 1944 book, *What is Life* [43], Schrödinger popularized Delbrück's idea and proposed that the gene could be regarded as an aperiodic crystal incorporating in its structure a submicroscopic molecular code specifying the development of the organism. Schrödinger also argued that in their ability to create "order out of disorder" living processes are not inconsistent with the second law of thermodynamics. Green plants effectively feed on the entropic decay of the sun as it generates the energy used in photosynthesis, and which is finally degraded to heat in the great mill-wheel of the global ecosystem [44]. Schrödinger's book had a significant impact in stimulating a number of young physicists to tackle the problems of heredity and variation. These physicists soon were to become the *avant garde* of the new biology. As Gunther Stent commented, "Schrödinger's book became a kind of *Uncle Tom's Cabin* of the revolution in biology that, when the dust had cleared, left molecular genetics as its legacy" [45].

It is ironic that in 1937, only two years after Delbrück's theory was published, Miroslav Demerec, working at Cold Spring Harbor, the nascent Mecca of molecular biology, found that mutation rates depend on various purely biological factors, including the particular gene used in the assay procedure, stage of the life cycle, and nutritional status of the organism. In addition, cold, as well as heat, was found to enhance mutation rates. Finally, in secret war-time research, Auerbach and Robson discovered the first (mustard gas) of the many hundreds of chemical mutagens that are now known to exist [46]. All of this led Muller, in 1952, to conclude that mutation is the result of "some biochemical disorganization in which processes *normally tending to hold mutation frequencies in check* are to some extent interfered with" [47]. The discovery in 1953 of the chemical structure of the gene by James Watson and Francis Crick [48] and, soon thereafter, of the enzymatic basis of its replication, recombination and repair, has made possible the formulation of a new, but complex, biochemical picture of genetic stability and change.

2   Biochemical Stabilization of the Genetic Material

Since "like begets like" in heredity, the genetic material must be replicated with the greatest precision from one generation of cells to the next. However, no real copying mechanism can work with perfect accuracy. The most skilful human key-punch operators make about one mistake per 5,000 characters. Even in mechanical printing presses, type faces break, inks smear, and we all know how well modern photographic copiers work! For continuing operation of even the most accurate machines repair is essential.

The number of accumulated errors in copying processes increases in proportion to the length of the message to be copied. Thus, the fidelity of DNA replication must be very great if lengthy chromosomes are to be reproduced accurately. Precise replication is essential both for the formation of normal progeny and the evolution of lengthy genetic messages [49].

Watson and Crick showed that the genetic material consists of two complementary strands of DNA linked side by side through rather weak hydrogen bonds. This double stranded structure takes the three-dimensional form of the famous "double helix" [48]. Each chain is a polymer composed of monomeric subunits called nucleotides. These monomers are made of sugar and phosphate moieties, which comprise the backbone of the strand, and one of four chemically distinct, organic bases linked to the sugars. These bases constitute the "letters" of the genetic code. There are two purine bases (adenine and guanine) and two pyrimidine bases (thymine and cytosine). The two strands of DNA are complementary because adenine in one strand should always be hydrogen-bonded with thymine in the other; similarly, guanine should be paired with cytosine. Sequences of approximately one to a few thousand bases in one strand typically encode the genetic message that constitutes a gene. Because of the complementary nature of the two DNA strands, the genetic information exists in redundant form within the genome. This redundancy provides the basis for DNA replication and certain important modes of DNA repair. These properties may explain, in part, the ubiquity of double-stranded DNA as the genetic material of organisms [50].

Replication of DNA is achieved by the unwinding of the double helix, accompanied by the polymerization of two new daughter strands each using one of the two parental strands as a complementary template. Expression of the genetic information encoded in the base sequence is accomplished in two steps called "transcription" and "translation." In the first step, the information corresponding to one (or a few) genes in DNA is copied by complementary polymerization into a single stranded "messenger" RNA (ribonucleic acid) molecule. In the second step, the sequence information in the messenger RNA molecule is translated into the corresponding amino acid sequence of the protein specified by the particular gene that was transcribed. The essence of the "central dogma" of molecular biology (DNA $\rightleftarrows$ RNA $\rightarrow$ protein) is that the flow of information from RNA to protein is one-way and cannot be reversed. This provides a molecular explanation for the non-inheritance of acquired characteristics: phenotypes (amino acid sequences of protein) cannot be translated back into genotypes (nucleotide sequences in RNA and DNA).

DNA replication is a complex biochemical process involving a variety of different enzymes [51]. Stripped to essentials, there are four basic requirements for the polymerization of nucleotides in DNA synthesis: first, an appropriate template from which the newly synthesized strand can be copied; second, the enzymes necessary to catalyze the various reactions involved; third, a suitable concentration of magnesium ions (necessary for optimal base pairing); and fourth, suitable concentrations of the four nucleotide precursors. A further battery of enzymes is necessary for synthesis of the nucleotides. These requirements are all rather stringent, and it would appear that if any fail to be met, the cell may die or become genetically unstable in the sense of exhibiting pathological levels of mutation and/or recombination [52]. Thus, mutation rates in cells may be increased by damage to DNA templates caused by various mutagenic chemicals or radiation, by substituting other metal ions for magnesium, and by altering the relative concentrations of the nucleotide precursors required for DNA synthesis. In addition, genetically heritable alterations in spontaneous mutation and recombination frequencies are observed in mutant organisms in which enzymes involved in DNA replication, repair, or nucleotide biosynthesis are structurally abnormal.

Normally in cells DNA replication is extremely accurate in accordance with the rules of complementary nucleotide base-pairing. When base mismatches do occur, these pairing errors can give rise to mutations in the next round of replication [53]. Unlike the error rate of one in five thousand for human key-punch operators, the error rate for DNA replication is about one mispaired nucleotide in every hundred million to ten billion replicated! On the other hand, non-enzymatic template-directed copying of polynucleotides has an estimated error rate of about one in a hundred [49]. Such a rate would not allow genetic messages to expand stably much beyond 100 nucleotides in length. The fantastic, though not perfect, accuracy of this process is achieved in bacteria through the combination of at least three mechanisms that actively promote the fidelity of DNA replication [54]. These are: first, the evolution of specific DNA polymerases that discriminate against the insertion of incorrect nucleotides at the point of polymerization in the nascent DNA strand. Such enzymes reduce the error rate from one in a hundred to about one in ten thousand. Second, the so-called editing or proof-reading function of DNA polymerases acts to catalyze the removal of mispaired nucleotides immediately after their insertion. This allows a further decrease in error rate to about one in ten million. Third, there exists a mismatch repair system capable of excising mispaired nucleotides in newly replicated, double-strand DNA. Although only a few replication errors escape these surveillance devices, mutations also

Table 1   DNA Damage and Its Repair in Typical Mammalian Cells*

Chemical Category	Damage Events per Hour	Repair Events per Hour (maximum)
1 Depurination	580	?
2 Depyrimidization	29	?
3 Deamination of cytosine	8	?
4 Single-strand breaks	2,300	200,000
5 Single-strand breaks subsequent to depurination	580	
6 Methylation of O^6-guanine	130	50,000
7 Pyrimidine dimers induced in human skin under noon Texas sun	50,000	50,000

*Data from R.R. Tice and R.B. Setlow, "DNA Repair and Replication in Aging Organisms and Cells," in *Handbook of the Biology of Aging*, ed. C.E. Finch and E.L. Schneider (2nd ed., New York: Van Nostrand Reinhold, 1985), 173–224.

It is clear from these data that a substantial amount of damage is produced spontaneously in DNA under normal physiological conditions. If DNA repair mechanisms did not exist, the accumulated damage would soon destroy the cell. Repair is necessary to prevent the spontaneous disintegration of the genetic information.

can arise as a result of damage caused by the physicochemical "decay" of DNA structure and by attack on DNA by various mutagenic agents.

The subunits of DNA – sugar, phosphate and the four organic bases – are all ordinary molecules. In themselves they present no features of special interest, philosophical or otherwise, to physicists and chemists. Deoxyribonucleic acid is a part and product of normal cellular metabolism. However, the very chemical vulgarity of DNA makes it prey to all the horrors and misfortune that might befall any molecule in a warm aqueous medium. Bits and pieces fall away, and even its backbone may be broken, at alarming rates under normal physiological conditions (Table 1). Even the immediate intracellular environment is not all that innocuous. It swarms with nasty items ever ready to chew upon genes and break their chemical bones. A partial list of these incubi would include certain highly reactive free radicals, peroxides, singlet oxygen, reducing agents, alkyl nitrosamines, nitrites, and alkylating agents, not to mention low, but sometimes significant exposures to ionizing radiation and ultraviolet light [55]. It is hardly surprising that, in addition to mechanisms that promote replicative fidelity, a complex battery of DNA repair systems has evolved to preserve the integrity of the genetic material against damage [6].

The existence of DNA repair was first postulated by radiation
biologists studying various modes of recovery of cells from the genetic
and lethal effects of ionizing and ultraviolet radiations. For when it
became clear that DNA was the principal target mediating such effects, it
was plausible to assume that these recovery phenomena were based
upon mechanisms for DNA repair [50]. The most seminal discovery in
this area was made by Richard Setlow and his colleagues at the Oak
Ridge National Laboratory in 1964. Setlow showed that bacteria
possess enzymes capable of excising from their DNA a specific class
of potentially lethal (and mutagenic) lesions induced by germicidal
ultraviolet radiation [56]. These lesions, formed by the photochemical
fusion of two adjacent pyrimidine bases in DNA, give rise to a local
distortion in the normal double helical structure of the genetic material.
Specific enzymes exist that are capable of "recognizing" such distortions and excising the damaged single-strand segment; the resulting
gap is then filled by "repair synthesis" using the opposite intact strand
as template for retrieval of the precise genetic information lost to
excision (see Figure 1 for a schematic diagram of this and other modes of
DNA repair). This excision repair process is capable of acting on DNA
damage produced by a large class of chemical mutagens in addition to
ultraviolet radiation. Mutants defective in this mode of repair are
extremely sensitive to these agents. It is indeed a rather general damage
detecting and correcting system [50]. However, it could not exist were
it not for the redundancy of the genetic information inherent in the
complementary double-stranded structure of DNA. Prior to the discovery of this mode of repair, the existence in cells of enzymes capable of
cutting and degrading the genetic material seemed paradoxical to
biochemists. However, during the past twenty years it has been found
that cells contain numerous distinct enzymes capable of cutting and
splicing DNA strands in all manner of highly specific ways. These
enzymes, many of which have been purified and are sold commercially,
now provide the basic "tool kit" for genetic engineers.

The excision repair system is capable of repairing DNA damage that
may be variously lethal, mutagenic, recombinagenic, or, in mammalian
cells, carcinogenic. People defective in such modes of repair are highly
susceptible to cancer [6]. Excision belongs to a class of processes
described operationally as "error-free repair" because no change in the
informational content of DNA results from their action. However, there
also exist certain so-called error-prone modes of DNA repair. These
systems actually generate mutations in the course of repairing or

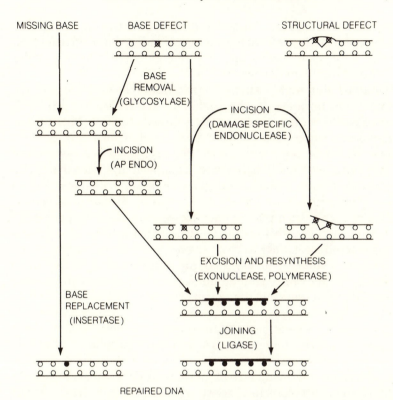

Figure 1 Schematic overview of the excision repair system. Three types of DNA lesions are illustrated (top): (a) missing bases (depurination or depyrimidiza-tion); (b) base defects (e.g., deaminated cytosine) unaccompanied by any significant change in the double-helical structure of DNA; and (c) structural defects accompanied by distortions in the DNA backbone (e.g., UV-induced pyrimidine dimers). In the classical scheme, the dimer-associated strand distortion is recognized by an endonuclease that cuts the backbone near the dimer (incision step). The lesion, along with varying numbers of adjacent nucleotides is excised and new DNA is synthesized in the resulting gap using the opposite intact strand as complementary template; repair is completed with a final ligation step. A second important mode of repair entails removal of the defective base (type b lesion) from the DNA backbone, followed by strand breakage and repair synthesis. Evidence also has been reported for the existence of enzymes ("insertases") that can directly replace missing bases, though doubt remains about the significance of these reports. (Diagram courtesy of Professor P.C. Hanawalt, Stanford University.)

bypassing DNA damage. Mutant cell lines defective in error-prone repair exhibit greatly *reduced* frequencies of induced mutagenesis [57]. Indeed, both spontaneous and induced mutation rates are under genetic and biochemical control.

The most thoroughly studied error-prone repair system is the so-called SOS response in the bacterium *Escherichia coli* [58]. This complex biochemical system normally is turned off in cells, but it may be turned on by exposure to suitable doses of ultraviolet light or other mutagens. It may seem paradoxical that cells possess a system that actually promotes mutational change. It has been suggested that such a system could serve to increase the chance of population survival, in highly mutagenic environments, by increasing the range of genetic variability within such populations. However, the SOS response also contributes significantly to cellular resistance to the lethal effects of the agents that induce it. On this basis, the response is consistent with the notion that evolution is blind to the future and that continued replication is the overwhelming immediate "priority" of the genetic process. It would seem that the slogan of cellular life is "better red than dead"!

DNA repair processes serve to reverse genetic damage after it has occurred. However, there also exist enzymes capable of neutralizing many common chemical mutagens, including those that arise as byproducts or intermediaries of normal oxidative metabolism. In addition, foodstuffs contain a variety of small molecules, which are not enzymes, but which function as antimutagens in protecting DNA against damage. These chemical and biochemical mechanisms constitute a first line of defence against many mutagenic substances of both endogenous and exogenous origin [55].

Many different enzymes, and hence many genes, are involved in the systems that promote replicative fidelity and protect DNA against damage. The actual number of such genes in any organism is not known, but it is likely to be rather large. In yeast, as many as one hundred genetic loci appear to be concerned with DNA repair [57]. In 1953, Dancoff and Quastler pointed out that if very great fidelity is to be achieved with equipment of poor precision, extensive checking procedures must be built into the system [59]. For optimum economy, the energy cost of such procedures should be just sufficient to reduce the error rate to a tolerable level. It seems likely that this "principle of maximum error" is exemplified in the genetic machinery of cells. The evolution of long genetic messages has been made possible by the fact that they encode extensive instructions for their own correction. The energy cost clearly is not prohibitive for the general economy of the cell, and the residual error rate is consistent both with the genetic integrity of the individual and the evolution of species.

The contemporary picture of the molecular basis of genetic stability and change is summarized in Figure 2. There exist many sources of DNA alteration or damage. If these had free rein, the mutation rate would be so great that the genetic integrity of cells could not be maintained: most would die of lethal mutations or erosion of the genome. Arrayed against these inevitable destabilizing influences is an amazing battery of biochemical mechanisms that promote genetic stability. However, being themselves composed of ordinary molecules, these systems cannot function with 100 per cent efficiency or perfect accuracy. The result of this Manichaean conflict between genetic order and decay is that the residual mutation rates (RMR) are extraordinarily low. However, even though many mutations are deleterious, natural selection has not, and indeed cannot, drive mutation rates to zero. Random genetic noise will be with us always, no matter how many ingenious mechanisms may evolve to combat it. Viewed in this light, the amazing diversity of life appears superimposed on mechanisms of genetic stability by the very noise those mechanisms were designed to suppress [49].

The diagram in Figure 2 illustrates the fact that mutationally altered phenotypes are macroscopic manifestations of chance events in the molecular domain. They arise from the thermodynamically ordained inability of the mechanisms for detoxification, replicational fidelity and DNA repair to suppress or reverse *all* genetic noise. That such effects, as in the pigmentation pattern of Siamese cats, can be seen in everyday life, is a tribute to the remarkable power of biochemical amplification inherent in the translation of genotype to phenotype through the genetic coding system [3]. With the aid of current DNA sequencing techniques, it is possible to identify the specific base alterations in the mutant genes involved. There is no other process known to physics in which such a dramatic macroscopic effect can be attributed to a chance atomic rearrangement in a single, *nameable* molecule. In addition to mutation, chance events are involved in evolution at all levels of biological organization, from the rearrangement of genomes to the extinction of species [61]. Thus, contemporary developments in molecular genetics and evolutionary theory, as well as in quantum mechanics, reveal a world that, in all its variety, owes as much to

Peroratio

The Darwinian mechanism of natural selection provides an observationally established explanation for adaptation and design in the biological world. Natural theology is a bankrupt enterprise. Indeed, ever since the criticisms of Hume and Kant, philosophers generally have concluded that the value of the Argument from Design "lies in its power to confirm belief, not create it, and in its power to set up a habit of mind, suitable for religious believers, in which there is a disposition to see design in the things about them" [62]. Some contemporary philosophers go so far as to assert that the design argument has no validity whatsoever and, even though it may serve to heighten religious emotion, is logically and morally indefensible.

As a result of research on the physical basis of evolution, it has become evident that the mutations which feed natural selection arise through chance events at the atomic and molecular levels. In the indeterminacy of their origin, at least some mutations resemble, metaphorically, the uncaused swerves of Epicurus' atoms in the void. Matter and life are veritably permeated with chance and accident. The basic order of nature is statistical in character, but these statistical laws enable us to make extremely accurate quantitative predictions of physical and chemical phenomena both as observed in the laboratory and in the world around us.

"Purposes" are conceived in the minds of sentient beings. Since no divine purposing agents can be shown objectively to exist, it seems reasonable to conclude that if there is any "purpose" of chance in nature it is to create the evolving variety of things that presumably could not come into existence in a completely deterministic universe. However, it is unlikely that such an imputed *post hoc* "purpose" will satisfy the psychic needs of people longing for some transcendent meaning in life. Nonetheless, it was upon a vision of fundamental indeterminacy in the atomic domain that Epicurus, in his Epistle to Menoeceus, developed his ethic of the prudent man who believes that "chance does not give man good and evil to make his life happy or miserable, but it does provide opportunities for great good or evil." For Epicurus "it would be better to accept the myth about the gods than to be a slave to the determinism of the [classical] physicists."

Recent developments in quantum cosmology suggest that the indeterminacy of individual atomic events may apply to the universe as a whole. The net energy of the universe seems to be close to zero, its total mass energy being balanced roughly by its negative gravitational potential energy. If the initial singularity of the "big bang" was a quantum fluctuation, we have before us a plausible scenario for the creation of the universe *in vacuo* [11]. On this basis, the macrocosm is as devoid of cause as is the spontaneous decay of any radioactive nucleus in the microcosm.

Subsequent to the "big bang," some current theories suggest that the universe "inflated" rapidly and developed a "foam-like" structure comprising a suitably astronomic number of "mini-universes," of which the cosmos that we observe is but one bubble in the foam. Within each of these mini-universes the properties of the elementary particles, and even the dimensionality of space-time, may be very different [11]. The extreme "fine tuning" of the physical constants of our universe may therefore indicate that only those universes possessing certain precise quantitative features will persist long enough, or have the required physical properties, to be observed as a result of the evolution of sentient beings capable of asking "why?" This implementation of the anthropic principle in quantum cosmology suggests a possible basis for cosmic design that is metaphorically related to the principle of natural selection: a large number of variant universes appear by chance, but only those with certain exceptional characteristics survive to be "selected," to become observable. However, the observability of such a world clearly does not entail its intelligibility, and other worlds that cannot be observed are best passed over in silence.

Since coherent, observationally plausible theories can account for design in both the physical and biological domains, science offers no reason to invoke the extra metaphysical baggage some would lay upon us with "god" hypotheses. Since gods and pure chance are both causally inexplicable, the physicist and priest W.G. Pollard has proposed an ultimate "god of the gaps" [63]. Pollard finds him lurking in those nooks and crannies of nature hidden from view by Heisenberg's uncertainty principle – the very places where Einstein sought his hidden variables. In a more biological vein, the Christian statistician D.J. Bartholomew has solemnly argued [64] that chance actually is "conducive to the kind of world which one would expect a God such as Christians believe in to create." He contends further that "chance was God's idea and he uses it to ensure the variety, resilience and freedom necessary to achieve his purposes" [64]. Perhaps then it is the Devil who really is invoked by those great merchants of chance in the insurance industry when they speak of "acts of God" and decline to pay our claims? The books of Pollard and Bartholomew, at the very least, demonstrate the proclivity of people to fashion gods, quite unashamedly, according to their prejudices, needs, and cultural presuppositions.

Anthropologists, sociologists, and psychologists have put forward cogent reasons for the ubiquitous development, and amazing elaboration, of religious ideas and practices in human, but presumably not simian, societies [65]. One reason why both sacred and secular religions exist is to ameliorate anxiety caused by our abililty to pose what seem to be deeply meaningful questions, but which, nonetheless, are unanswerable or meaningless. The universal need to develop belief systems, or myths, that suppress the uncertainties of self-conscious life, and offer a coherent account for existence, might itself be an adaptation of cultural evolution that ensures the stability of brain and mind in *Homo sapiens*. From this point of view, myth-making contributes to the health and survival of the individual and the species [7]. The interminable rumination and cosmic speculations of certain obsessive neurotics are well known to psychiatry. Physical pain leads to the release of endorphins, cerebral opiates that simultaneously remove the pain and deflect attention from its source. Perhaps through analogous mechanisms, mythologies and religious exercises dampen anxiety stimulated in confrontations with the mysterious. There may be more truth than poetry in the metaphor that theism is the opiate of the masses, that Marxism is the opiate of intellectuals, and that Darwinism is the opiate of scientists.

The essence of "isms" is that they offer simplistic solutions to what are in fact very complex, even intractable, problems. Uncritical belief in the truth of such systems breeds arrogance at best, fanaticism at worst. But we have good reason to be humble because our brains are so constructed that limits exist to the scientific understanding of man and of nature [66]. Even J.B.S. Haldane, in a rare moment of modesty, once commented that "the universe is not only queerer than we supppose, it is queerer than we *can* suppose." For Kirkegaard, the "supreme paradox of all thought is the attempt to discover something that thought cannot think." It seems to me that art alone provides genuine solace for the soul, and brings us into contact with the humanity that lies within and among us all. In the words of W.H. Auden, "without art we should have no notion of the sacred; without science we should always worship false gods."

I thank the many colleagues who were kind enough to read this paper and offer helpful criticisms, especially F.J. Ayala, P.R.J. Burch, J.F. Crow, J.W. Drake, G.B. Golding, T.H. Jukes, J.G. Little, L.A. Loeb, R.A. Morton, P.J.E. Peebles, and S. Wolff.

NOTES AND REFERENCES

1 Epicurus extended the atomism of Democritus to include the idea of un-caused (pure chance) "swerves" of the atoms as they fall in the void. He considered that the resulting collisions of different kinds of atoms produced the varieties of compound bodies seen in nature, and allowed also for human freedom of will. Epicurean ethics are based upon this indeter-ministic atomism. In the words of Lucretius, whose poem *De Rerum Natura*, contains the most complete, extant presentation of Epicurean philosophy:

> The atoms, as their own weight bears them down
> Plumb through the void, at scarce determined times,
> In scarce determined places, from their course
> Decline a little – call it, so to speak,
> Mere changed trend. For were it not their wont
> Thuswise to swerve, down would they fall, each one,
> Like drops of rain, through the unbottomed void;
> And then collisions ne'er could be nor blows
> Among the primal elements; and thus
> Nature would never have created aught.

(Lucretius, *On the Nature of Things*, verse translation by W.E. Leonard [Los Angeles: The Limited Editions Club, 1957]).

The classic description of Epicureanism is given by Cyril Bailey, *The Greek Atomists and Epicurus* (Oxford: Clarendon Press, 1928). The few extant writings of Epicurus himself may be found in R.M. Geer, *Letters, Principal Doctrines and Vatican Sayings of Epicurus* (Indianapolis: Bobbs-Merrill, 1964). A favourable appraisal of Epicureanism is provided by B. Farrington, *The Faith of Epicurus* (London: Weidenfeld and Nicolson, 1967).

2 I. Newton, *Mathematical Principles of Natural Philosophy and his System of the World*, trans. A. Motte, revised by F. Cajori (Berkeley: University of California Press, 1960), 546.

3 T.F. Smith and H.J. Morowitz, "Between History and Physics," *J. Mol. Evol.* 18 (1982), 265–82. See also M. Eigen and R. Winkler, *The Laws of the Game* (New York: Knopf, 1981).

4 C.R. Darwin and A.R. Wallace, "On the Tendency of Species to Form Varieties; and on the Perpetuation of Varieties and Species by Natural Means of Selection," *J. Proc. Linn. Soc. (Zoology)* 3 (1859): 45–62.

5 C.R. Darwin, *On the Origin of Species by Means of Natural Selection* (London: John Murray, 1859), 82.

6 E.C. Friedberg, *DNA Repair* (New York: W.H. Freeman, 1985).

7 F. Jacob, *The Possible and the Actual* (Seattle: University of Washington Press, 1982).

8 N. Bohr, *Atomic Theory and the Description of Nature* (New York: Macmillan, 1934). Recent experiments in Paris, carried out by A. Aspect and his colleagues, finally seem to have vindicated the Copenhagen interpretation of quantum theory and ruled out any form of local "hidden variables" theory. For a general discussion of this work, see F. Rohrlich, "Facing Quantum Mechanical Reality," *Science* 221 (1983): 1251–55.

9 I. Prigogine, *From Being to Becoming* (San Francisco: W.H. Freeman, 1980).

10 J. Monod, *Chance and Necessity* (New York: Knopf, 1971).

11 The original notion of "quantum cosmology" was put forward by E.P. Tryon in "Is the Universe a Vacuum Fluctuation," *Nature* 246 (1973): 396–97. In this brief paper Tryon wryly remarked that he offered "the modest proposal that our Universe is simply one of those things which happen from time to time." The idea, based on the plausible assumption that the net energy of the universe is zero, has stimulated much further research by eminent cosmologists. For a recent technical review see A.D. Linde, "The Inflationary Universe," *Rep. Prog. Phys.* 47 (1984): 925–86.

12 M. Delbrück, *Mind from Matter?* G.S. Stent et al. (Palo Alto: Blackwell Scientific Publications, 1986).

13 F. Jenkin, "The Origin of Species," *North Brit. Rev.* 46 (1867): 277–318: reprinted in D.L. Hull, *Darwin and His Critics* (Cambridge, Mass: Harvard University Press, 1973).

14 S. Butler, *Evolution, Old and New* (London: Hardwicke and Bogue, 1879), 363.

15 C.R. Darwin, *The Variation of Animals and Plants under Domestication* (London: John Murray, 1868), 2: 357.

16 It is a curious historical twist of mid-twentieth century science that bacteria, the very organisms that provided experimental material for the rise of modern molecular genetics, also offered the last sanctuary for Lamarckism in biology. The appearance of bacterial resistance to attack by viruses, and also to antibiotic drugs, was considered to be an immediate, heritable adaptation brought about by exposure to these agents. The demonstration, by Luria and Delbrück, and the Lederbergs, that the resistant cells arise by spontaneous mutation in cultures never previously exposed to such agents, not only provided a Darwinian interpretation of the phenomenon, but also laid the foundation for the subsequent dramatic development of bacterial genetics. See S.E. Luria and M. Delbrück, "Mutations of Bacteria from Virus Sensitivity to Virus Resistance," *Genetics* 28 (1943): 491; J. Lederberg and E.M. Lederberg, "Replica Plating and Indirect Selection of Bacterial Mutants," *J. Bacteriol.* 63 (1952): 399.

17 E.R. Lankester, "Zoology," in *Encyclopaedia Britannica* (9th ed. Boston: Little, Brown, 1875–89), 24: 806.

18 D.W. Thompson, *On Growth and Form* (Cambridge: Cambridge University Press, 1917).

19 S.J. Gould and R.C. Lewontin, "The Spandrels of San Marco and the Panglossian Paradigm: A Critique of the Adaptationist Programme," *Proc. R. Soc. Lond.* B205 (1979): 581–98.

20 G. Mendel, "Versuche über Pflanzen-Hybriden," *Verhand. Naturforsch. Ver. Brünn* 4, (1865): 3–47; English translation by E.R. Sherwood in *The Origin of Genetics*, ed. C. Stern and E.R. Sherwood (San Francisco: W.H. Freeman, 1966).

21 W. Bateson, *Mendel's Principles of Heredity* (Cambridge: Cambridge University Press, 1909).

22 T.H. Morgan, A.H. Sturtevant, H.J. Muller, and C.B. Bridges, *The Mechanism of Mendelian Heredity* (New York: Henry Holt, 1915).

23 S.S. Chetverikov, "On Certain Aspects of the Evolutionary Process from the Standpoint of Modern Genetics," *Proc. Am. Phil. Soc.* 105 (1961): 167–95; an English translation by M. Barker, ed. I.M. Lerner, of the original Russian text (*Zh. Eksp. Biol. Med.* A2 [1926]: 3–54).

24 R.A. Fisher, *The Genetical Theory of Natural Selection* (Oxford: Clarendon Press, 1930).

25 J.B.S. Haldane, *The Causes of Evolution* (New York: Harper, 1932).

26 S. Wright, "Evolution in Mendelian Populations," *Genetics* 16 (1931): 97–159.

27 T. Dobzhansky, *Genetics and the Origin of Species* (New York: Columbia University Press, 1937).

28 E. Mayr, *The Growth of Biological Thought* (Cambridge, Mass: Harvard University Press, 1982).

29 R.C. Lewontin, *The Genetic Basis of Evolutionary Change* (New York: Columbia University Press, 1974).

30 T. Dobzhansky, *Genetics of the Evolutionary Process* (New York: Columbia University Press, 1970).

31 M. Kimura, *The Neutral Theory of Molecular Evolution* (Cambridge: Cambridge University Press, 1983).

32 M. Nei and R.K. Koehn, eds., *Evolution of Genes and Proteins* (Sunderland, Mass: Sinauer Associates, 1983).

33 G.L. Stebbins and F.J. Ayala, "Is a New Evolutionary Synthesis Necessary?" *Science* 213 (1981): 967–71; also "The Evolution of Darwinism," *Scientific American* 253 (1985): 72–82.

34 J.F. Crow and C. Denniston, "Mutation in Human Populations," in *Advances in Human Genetics*, ed. H. Harris and K. Hirschorn (New York: Plenum Publishing, 1985), 59–123; see also "The Mutation Component of Genetic Damage," *Science* 212 (1981): 888–93.

35 S. Ohno, *Evolution by Gene Duplication* (New York: Springer-Verlag, 1970).

36 W. Gilbert, "Genes-in-Pieces Revisited," *Science* 228 (1985): 823–4.

37 G. Dover, "Molecular Drive: A Cohesive Mode of Species Evolution," *Nature* 299 (1982): 111–17.

38 A. Campbell, "Evolutionary Significance of Accessory DNA Elements in Bacteria," *Ann. Rev. Microbiol.* 35 (1981): 55–83.

39 J.D. Watson, J. Tooze and D.T. Kurtz, *Recombinant DNA* (New York: Scientific American Books, 1983).

40 H.J. Muller and E. Altenburg, "The Rate of Change of Hereditary Factors in *Drosophila*," *Proc. Soc. Exptl. Biol. Med.* 17 (1919): 10–14.

41 H.J. Muller, "Artificial Transmutation of the Gene," *Science* 66 (1927): 84–7; see also L.J. Stadler, "The Comparison of Ultraviolet and X-ray Effects on Mutation," *Cold Spring Harbor Symp. Quant. Biol.* 9 (1941): 108–77.

42 N.W. Timoféeff-Ressovsky, K.G. Zimmer, and M. Delbrück, "Über die Natur den Genmutation und der Genstrucktur," *Nachr. a.d. Biologie d. Ges. d. Wiss. Göttingen* 1 (1935): 189–245. Delbrück's theory of the physical basis for gene stability is contained in Part III of this paper.

43 E. Schrödinger, *What is Life?* (Cambridge: Cambridge University Press, 1944).

44 H.J. Morowitz, *Energy Flow in Biology* (New York: Academic Press, 1968).

45 G.S. Stent, *The Coming of the Golden Age* (New York: Natural History Press, 1969).

46 C. Auerbach and J.M. Robson, "The Production of Mutations by Chemical Substances," *Proc. R. Soc. Edinburgh,* B62 (1947): 271–283; see also C. Auerbach, *Mutation Research* (London: Chapman and Hall, 1976).

47 H.J. Muller, "The Nature of the Genetic Effects Produced by Radiation," in *Radiation Biology*, ed. A. Hollaender (New York: McGraw-Hill, 1954–6) 1: 417.

48 J.D. Watson and F.H.C. Crick, "A Structure for Deoxyribonucleic Acid," *Nature* 171 (1953): 737–8; and "Genetical Implications of the Structure of Deoxyribonucleic Acid," *Nature* 171 (1953): 964–7. For authoritative texts on the principles of molecular biology, see J.D. Watson, *Molecular Biology of the Gene* (3rd ed. Menlo Park, Calif: W.A. Benjamin, 1976), or D. Freifelder, *Molecular Biology* (Boston: Science Books International,

1983). Contemporary popular accounts of many of the most notable developments in molecular biology, written by the scientists directly involved, are collected in *The Molecular Basis of Life: Readings from "Scientific American,"* ed. with introductions R.H. Haynes and P.C. Hanawalt (San Francisco: W.H. Freeman, 1968).

49 D.C. Reanney, D.G. MacPhee, and J. Pressing, "Intrinsic Noise and the Design of the Genetic Machinery," *Aust. J. Biol. Sci.* 36 (1983): 77–91; see also M. Eigen and P. Schuster, *The Hypercycle: A Principle of Natural Self-Organization* (Berlin: Springer-Verlag, 1979).

50 P.C. Hanawalt and R.H. Haynes, "The Repair of DNA," *Scientific American* 216 (1967): 36–43.

51 A. Kornberg, *DNA Synthesis* (San Francisco: W.H. Freeman, 1974).

52 R.H. Haynes, "Molecular Mechanisms in Genetic Stability and Change: Role of Deoxyribonucleotide Pool Balance," in *Genetic Consequences of Nucleotide Pool Imbalance*, ed. F. de Serres (New York: Plenum Publishing, 1985).

53 J.W. Drake and R.H. Baltz, "The Biochemistry of Mutagenesis," *Ann. Rev. Biochem.* 45 (1976): 11–37.

54 L.A. Loeb and T.A. Kunkel, "Fidelity of DNA Synthesis," *Ann. Rev. Biochem.* 52 (1982): 429–457.

55 B.N. Ames, "Dietary Carcinogens and Anticarcinogens," *Science* 221 (1983): 1256–64.

56 R.B. Setlow and W.L. Carrier, "The Disappearance of Thymine Dimers from DNA: An Error-Correcting Mechanism," *Proc. Natl. Acad. Sci. U.S.A.* 51 (1964): 226–31.

57 R.H. Haynes and B.A. Kunz, "DNA Repair and Mutagenesis in Yeast," in *The Molecular Biology of the Yeast Saccharomyces: Life Cycle and Inheritance*, ed. J. Strathern, J.R. Broach, and E.W. Jones (New York: Cold Spring Harbor Laboratory, 1981), 371-414. The concept of error-prone repair was developed by E.M. Witkin, "Ultraviolet-induced Mutation and DNA Repair," *Ann. Rev. Genet.* 3 (1969): 525–52.

58 G. Walker, "Mutagenesis and Inducible Responses to DNA Damage in *Escherichia coli*," *Microbiol. Rev.* 48 (1984): 60–93.

59 S.M. Dancoff and H. Quastler, "The Information Content and Error Rate of Living Things," in *Information Theory in Biology*, ed. H. Quastler (Urbana: University of Illinois Press, 1953), 263–73.

60 J.W. Drake, "The Role of Mutation in Microbial Evolution," *Symp. Soc. Gen. Microbiol.* 24 (1974): 41–58.

61 S.J. Gould, "Chance Riches," rep. in his *Hen's Teeth and Horse's Toes* (New York: Norton, 1983), 332–42.

62 T. McPherson, *The Argument from Design* (London: Macmillan, St. Martin's Press, 1972); see also J.L. Mackie, *The Miracle of Theism* (Oxford: Clarendon Press, 1982).

63 W.G. Pollard, *Chance and Providence* (New York: Scribner, 1958).

64 D.J. Bartholomew, *God of Chance* (London: SCM Press, 1984).

65 J.M. Yinger, *The Scientific Study of Religion* (New York: The Macmillan Company, 1970).

66 G.S. Stent, "Limits to the Scientific Understanding of Man," *Science* 187 (1975): 1052–7. See also his *Paradoxes of Progress* (San Francisco: W.H. Freeman and Company, 1978) and *Morality as a Biological Phenomenon* (Berkeley: University of California Press, 1980).

BARRY W. GLICKMAN

"The Gene Seemed as Inaccessible as the Materials of the Galaxies"

The sciences in our century have been marked almost everywhere by momentous discoveries, by extraordinary people, by revolutions in understanding, by a remarkable dynamism of legions of scientists in a world-wide community. Twice since the early 1900s science has generated a transformation so broad and so deep that it touches our most intimate sense of the nature of things. The first of these transformations was in physics, the second in biology.

The revolution in physics came first. It began with quantum theory and the theory of relativity, with intellectual giants such as Max Planck and Albert Einstein, and came to encompass both the interior of the atom and the structure of space and time. This giant step forward had altered modern quantum mechanics by the mid-1930s. Most of what has happened in physics since then, at least until most recently, has been the playing out of these great themes.

Biology stands in contrast. In molecular biology, still in its infancy in the 1930s, an outline of the nature of life emerged only around 1970. Since then, with the use of the new tools of genetic manipulation, this outline is beginning to fill out. The understanding of the nature of DNA, the so-called stuff of life, is becoming more refined. Details of the fidelity of its replication, the mechanisms of its repair, are becoming clear. We are now beginning to gain an insight into the conservation of genetic material. We understand better the transfer of the genetic information held in DNA to the proteins that provide the structures and functions of all living matter. And we are beginning to appreciate the details of gene structure and gene regulation that not only permit sophisticated life forms but also account for the differences between a human being and a mouse and, indeed, between men and women, this latter endearing at least some aspects of biology to the general public.

I shall present a brief history of molecular biology, outlining our current perceptions of what genes are and how they function. I shall indicate how cells have evolved to conserve their genetic heritage and discuss our current view of how mutations arise. Finally, I shall attempt to describe the future of this modern science; its consequences are likely to be felt by all mankind.

Origins of Molecular Biology

Modern biology, as we know it, had its origins in the early nineteenth century. The development, in Germany, of powerful light microscopes and techniques for fixing and staining living tissues led to the generalization that all living material, whether of plant or animal origin, was made up of tiny box-like entities called cells. The cell became known as the fundamental unit of life. By the 1860s rod-like bodies within the nucleus were identified. These bodies stained strongly with certain basic dyes and hence were called chromosomes. As the movement of the chromosomes during cell division was recognized in the late 1880s, it became clear that they possessed the expected behaviour of cellular elements postulated to be responsible for the passing of genetic traits from one generation to the next. Proof for the idea that chromosomes are the bearers of heredity could only come after the rediscovery in 1901 of the genetic experiments of George Mendel. The first trait assigned to a chromosome location was sex itself, in 1905. It was natural to speculate that all traits might be chromosomally located. The first confirmation came soon from genetic crosses with eye colour mutants of the fruit fly *Drosophila*, by T.H. Morgan [1]. In the period prior to the First World War, numerous mutants were mapped to both the sex and non-sex (or autosomal) chromosomes. By the end of the 1920s H. Muller [2] and L. Stadler [3] had independently discovered that X-rays induced mutations and thereby provided geneticists with a way of obtaining large numbers of mutants. By the late 1930s evidence for at least 1500 genes on the four chromosomes of *Drosophila* had accumulated.

Biochemistry provides the next link in our story. Although initially much controversy surrounded the chemical identity of proteins, as early as 1880 there were hints that at least some enzymes were proteins. But it was not until the 1930s that the protein nature of all enzymes was universally accepted. Nevertheless, long before the essence of proteins was found to reside in their amino acid sequences, scientists speculated that genes somehow controlled the synthesis of specific enzymes. The concept was first clearly proposed by the English physician Archibald Garrod, who studied what he called "inborn errors" in metabolism. In the genetic disease phenylketonuria, the amino acid phenylalanine cannot be converted to the related amino acid tyrosine. Garrod correctly hypothesized that this biochemical defect arose because a mutant gene failed to provide the correct enzyme. Because enzymes are always proteins, the concept would come to be stated as "one-gene-one-protein." Evidence for this formulation came in the early 1950s when the chemist Linus Pauling demonstrated that the sicle-cell anemia was due to altered hemoglobin molecules that have a reduced affinity for oxygen.

The Nature of Genetic Material

By the mid-1920s it had been established that DNA was found only in the chromosomes and hence had the location expected for genetic material. Histones, the second major component of chromosomes, could be ruled out as genetic material because they are absent from most sperm cells, where they are replaced by protamines. Yet scientists were reticent about accepting DNA as genetic material. They thought that this simple molecule composed of just four bases could not account for the specificity of heredity. It was commonly held that some as yet uncharacterized protein was the true genetic material. The first hint that DNA was indeed the genetic material was the result of a fortuitous observation in 1928 by the Englishman Frederick Griffith, who observed that non-virulent strains of pneumococcus bacteria could be made virulent by exposing them to virulent bacteria killed by high temperature. This discovery was confirmed and extended by Avery and his colleagues [4]. They fractionated the dead cells and demonstrated that it was not protein, nor RNA, the other form of nucleic acid, but DNA that was responsible for the transformation. Their results were published in 1944.

Still, there was tremendous reluctance to accept DNA as the true genetic material. Many sceptics preferred to believe that Avery had missed finding the "genetic protein." At this point the characterization of viruses as discrete particles made up of DNA and protein paved the way for the definitive experiments. The question was finally settled in 1952, when at Cold Spring Harbor Alfred Hershey and Martha Chase

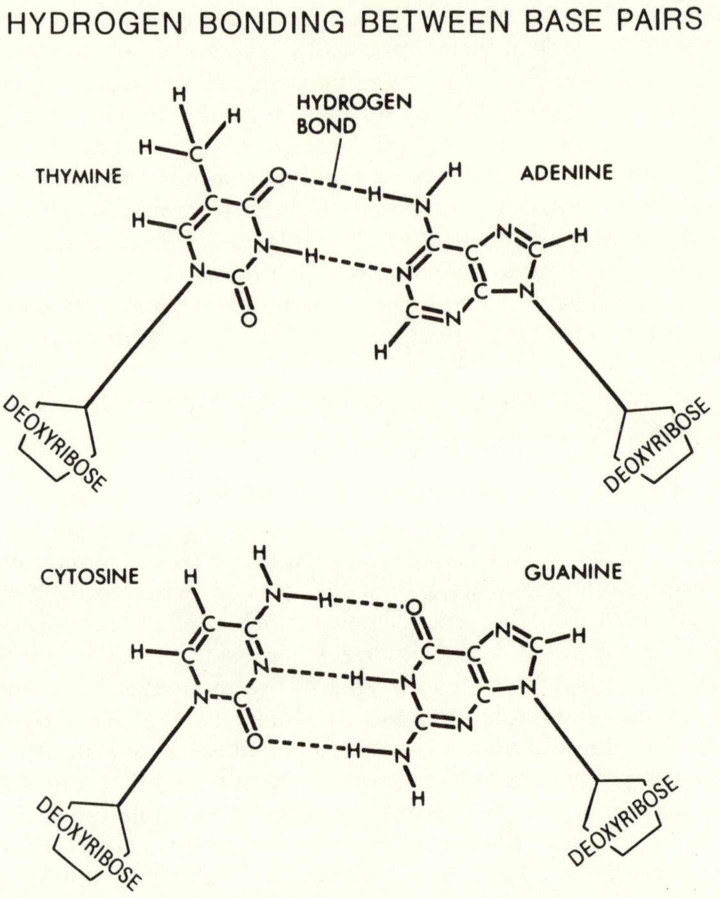

HYDROGEN BONDING BETWEEN BASE PAIRS

Figure 1

[5] showed that only the DNA of bacterial viruses entered the host. The protein coats remained outside the cells and hence could be ruled out as carriers of genetic information.

Characterization of DNA

High-level crystallographic analysis of DNA began seriously in 1949 in the physics department of King's College, London. Maurice Wilkins and Rosalind Franklin soon obtained remarkably detailed X-ray pictures of DNA. The X-ray photographs revealed a highly regular

BASE PAIRING OF TWO DNA CHAINS

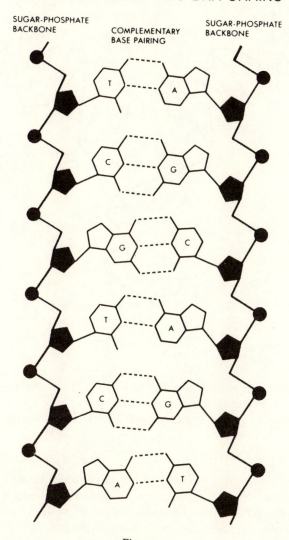

Figure 2

molecule in which two or more strands were intertwined. In the spring of 1952 the structure of DNA was resolved by James Watson and Francis Crick [6]. They proposed that DNA was composed of two polynucleotide chains running in opposite directions and held together by

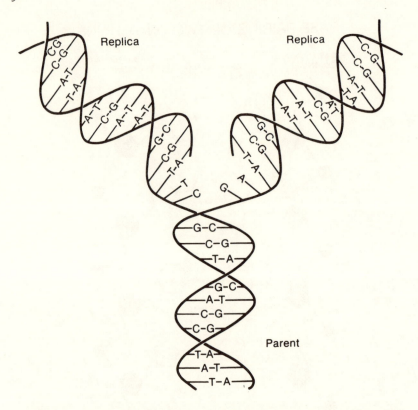

Figure 3 The self-complementarity of DNA provides a mechanism for its own duplication. While the enzymology is complex, the basic notion of how one DNA molecule can be precisely duplicated became clear when the structure of DNA had been elucidated.

hydrogen-bonding between the pairs of centrally located bases, the "double helix." In the helix adenine (A) is always hydrogen-bonded to thymine (T) and guanine (G) is always paired with cytosine (C). The hydrogen-bonding between the bases present in DNA is illustrated in Figure 1. Although the hydrogen-bonding between base pairs is weak, the long chains of DNA permit the formation of a stable, double-stranded structure under normal physiological conditions. The double-stranded nature of DNA is diagrammed in Figure 2.

Before the structure of DNA was known, it was exceedingly difficult to speculate on how genes were accurately duplicated. The intrinsic

ORDER OF GENE MUTATIONS IS THE SAME AS THAT OF AMINO-ACID CHANGES

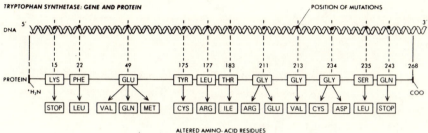

ALTERED AMINO-ACID RESIDUES

Figure 4 The colinearity of a gene is illustrated by the fact that the position of mutations in the DNA corresponds directly to the positions of alterations in amino acid sequence.

serendipity of the double helix is that it divined its own function. What Watson and Crick realized was that the two intertwined strands of DNA possessed the molecular complementarity needed to direct its own accurate replication [7]. The DNA helix can be regarded as a pair of positive and negative templates. Each strand is capable of specifying its complement. Hence, as shown in Figure 3, the double helix is capable of templating the synthesis of two daughter strands!

Development of Fine Structure Genetics

The great speed with which the basic facts of molecular genetics emerged following the discovery of the double helix was made possible by a collective decision made in the mid-1940s to focus whenever possible on the bacterium *Escherichia coli* and its bacteriophage. Soon the gene could be defined as a collection of adjacent nucleotides that specify the amino acid sequences of the cellular proteins. Simplicity argued that the corresponding nucleotide and amino acid sequences would be colinear. This hypothesis was quickly confirmed by correlating the relative positions of mutations in a gene with changes in its polypeptide product. The best early data on the colinearity of the gene (see Fig. 4) were obtained by Charles Yanofsky [8], who studied a gene coding for an enzyme involved in the synthesis of the amino acid tryptophan. He demonstrated rather convincingly that the relative order of each amino acid replacement in the protein was the same as its respective mutation on the genetic map.

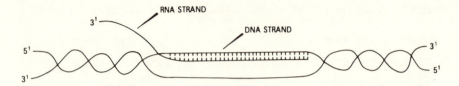

Figure 5 This artistic impression shows how the genetic information stored in the DNA can be transferred to messenger RNA. The messenger can then be released to transport the information stored in DNA to the cytoplasm where the synthesis of proteins takes place.

A direct role for DNA in templating the amino acid sequences of protein is not possible because DNA is located in chromosomes sequestered in the nucleus while most protein synthesis occurs in the cytoplasm. The genetic information of the DNA must then be transferred to an intermediate which then moves into the cytoplasm where it directs the synthesis of proteins. It was known that cytoplasm actively engaged in protein synthesis contained large amounts of RNA. Moreover, RNA is very similar to DNA. It has a slightly different sugar in its sugar-phosphate backbone and uracil in place of thymine, but it can be readily shown to be complementary to DNA. Hence it is quite reasonable to conceive of mechanisms as in Figure 5, by which DNA can template the production of RNA. This RNA would then migrate to the cytoplasm and direct the synthesis of specified proteins. These RNAS have therefore become known as messenger or mRNA molecules.

The ultimately correct relationship between DNA, RNA, and protein was brilliantly conceived by Francis Crick in 1953. To quote him directly: "My mind was, that a dogma was an idea for which there was no reasonable evidence! You see?" According to this dogma, for which there was no direct evidence, and which was a subject of speculation for some and perhaps faith for others: the genetic material was DNA; it could direct its own accurate replication; it was the template for RNA production and that RNA directed the synthesis of proteins. In essence, what is now known as the Central Dogma of Molecular Biology states that information can pass from nucleic acid to nucleic acid and from nucleic acid to protein, but never out again. This has held true. We know of no case where the informational transfer has involved the translation of information from the language of the proteins into that of the nucleic acids. Crick's declaration of the Central Dogma has only been updated: we now also know that RNA can be translated into DNA

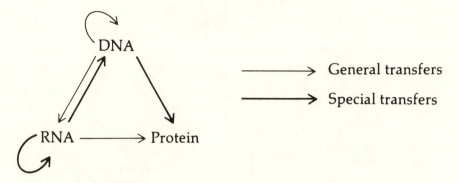

Figure 6 The Central Dogma states which informational transfers can occur. No case of the translation of information contained in protein to the language of the nucleic acid is known.

as in the case of reverse transcriptase directing the synthesis of DNA on the template supplied by viruses whose extracellular genetic soul consists of RNA. The principles of the Central Dogma can be illustrated graphically as in Figure 6.

With respect to the title of this conference it seems particularly appropriate to point out the wickedness of Crick's humour in announcing the Central Dogma. That is that the word "DOGMA" in reverse is "AM GOD"!

Roles of Templates, Enzymes, and tRNA in Protein Synthesis

Though originally it was speculated that there was some topological property by which RNA might directly template the fitting of amino acids, it was quickly recognized that this was very unlikely. This dilemma was solved by Crick, who proposed that adapter molecules exist to which amino acids can bind but which are also complementary to specific groups of nucleotides, each group, or codon, coding for a specific amino acid. Testing this hypothesis, however, had to wait for the development of *in vitro* methods for the analysis of RNA and protein synthesis. Thus, while the discovery of the double helix was a cause for pure delight to geneticists, the task of the biochemist was just beginning.

By 1960 both DNA and RNA had been synthesized in highly purified cell-free extracts. The biochemical nature of both RNA and DNA had been resolved and the models of Watson and Crick largely proven. In

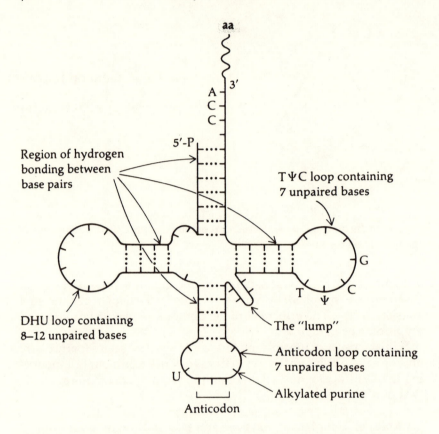

Figure 7 A general model for a tRNA molecule. These tiny molecules are essential components of the protein synthesis machinery of the cell. Each tRNA specifies a specific amino acid. One end of the molecule becomes the bearer of an amino acid while the other end contains the "anticodon" which is complementary to the "codon" specifying the incorporation of the specific amino acid.

terms of protein synthesis, however, many preconceived notions had to be discarded. It was discovered, for example, that while the cytoplasmic particles called ribosomes are the site for protein synthesis, the RNA contained in the ribosomes serves no templating function but rather has structural importance. Indeed, it emerged that only a minor fraction, only 2 to 5 per cent, of the cellular RNA served templating functions.

THE GENETIC CODE

First position (5' end)	Second position				Third position (3' end)
	U	C	A	G	
U	Phe	Ser	Tyr	Cys	U
	Phe	Ser	Tyr	Cys	C
	Leu	Ser	Stop	Stop	A
	Leu	Ser	Stop	Trp	G
C	Leu	Pro	His	Arg	U
	Leu	Pro	His	Arg	C
	Leu	Pro	Gln	Arg	A
	Leu	Pro	Gln	Arg	G
A	Ile	Thr	Asn	Ser	U
	Ile	Thr	Asn	Ser	C
	Ile	Thr	Lys	Arg	A
	Met	Thr	Lys	Arg	G
G	Val	Ala	Asp	Gly	U
	Val	Ala	Asp	Gly	C
	Val	Ala	Glu	Gly	A
	Val	Ala	Glu	Gly	G

Figure 8 The genetic code is summarized. The codon AUG uniquely specifies the incorporation of Methionine while CCU, CCC, CCA, and CCG all specify the incorporation of Proline.

Equally important was the discovery that prior to their incorporation into proteins, the amino acids are chemically linked to small RNA called transfer or tRNA. The amino-acid bearing tRNA molecules recognize the groups of nucleotides that code for their respective amino acids and line up onto the mRNA. No direct contact exists between the amino acids and mRNA. Hence the tRNA molecules are the "adapter molecules" that had been predicted several years earlier by Crick. The general structure of a tRNA molecule is shown in Figure 7.

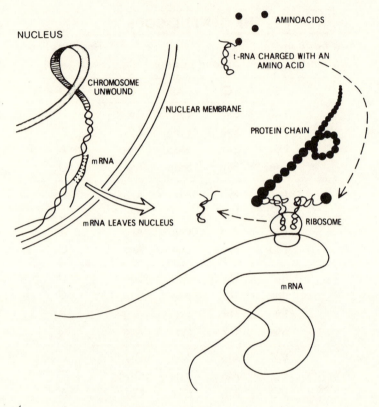

Figure 9 The details of protein synthesis are summarized. The DNA normally packed tightly in the chromosomes becomes unwound and accessible to RNA polymerases so that a messenger or mRNA molecule is produced. The mRNA then travels to the cytoplasm where it uses the structural support of the ribosome to direct the incorporation of specific amino acids into the protein coded. The amino acids for protein synthesis are supplied by tRNA molecules which transport specific amino acids to an appropriate site on the ribosome. The process of adding amino acids to the protein continues until the protein is complete. At this time the protein is released from the ribosome.

The Genetic Code

An exhaustive study of bacteriophage T4 mutants that contained additions or deletions of a single base pair led Sydney Brenner and Francis Crick to the conclusion that each codon was composed of three

REPRESSED STATE

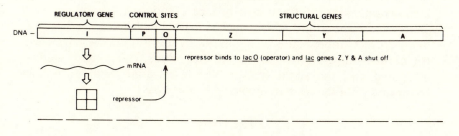

INDUCED STATE

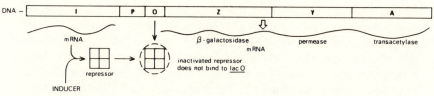

Figure 10 The *lac* operon is diagrammed. The structural genes are coded by a single, multi-cistronic mRNA. In the absence of lactose (or another inducer), the repressor normally binds to a region of the control elements and blocks the expression of the operon. This is known as the "repressed" state. In the presence of lactose or another inducer, the repressor becomes inactivated and the genes of the operan are expressed.

nucleotides. Using the four nucleotides in groups of three, there was the potential to code for sixty-four amino acids. Taking advantage of synthetic mRNA molecules of known composition, the elucidation of the genetic code was soon complete. It is presented in Figure 8. Only sixty-one of the sixty-four possible combinations of three code for specific amino acids. Three of the sixty-four possible codons do not specify a specific amino acid, but rather are used as punctuation and found to function as stop signals for protein elongation. It was soon recognized that there was a preferred start codon, AUG, which specifies methionine. The fact that more than one codon codes for the same amino acid is denoted as the degeneracy of the code. This degeneracy serves as a great buffer, however, since many mutations in DNA sequence are "silent," or do not result in an alteration of the protein sequence.

The mechanism of protein synthesis is now relatively well understood. The basics of the model are illustrated in Figure 9. The ribosome serves as a scaffold, a physical structure for protein synthesis. The mRNA supplies the information in the form of codons specifying specific amino acids. The amino acids, attached to tRNA molecules, are directed to the correct binding sites on the mRNA by the anticodons, which are triplet nucleotides complementary to the codons. The protein chain is elongated and the spent tRNA molecule released. The process is continued until the protein molecule is completed.

The Gene

We have now reached the stage where we can begin to consider a gene as not only a unit of heredity, but also a functional entity. We understand the language. We know how the amino acids that make up the protein are encoded and how this information is translated into the protein. As one might expect, the expression of genes is regulated. Numerous mechanisms for the expression and repression of genes have been identified. One of the simplest cases, that of the *lac* operon of *Escherichia coli*, is diagrammed in Figure 10. A series of functionally related genes make up what is known as an operon. This particular operon encodes a few genes required for the metabolism of the sugar, lactose. When the cell is grown without lactose a protein called a repressor binds to a regulatory sequence known as the operator. When the cell is grown in lactose and requires the enzymes for fermenting this sugar, the repressor protein is specifically inactivated and the operator becomes functional. As a consequence, the mRNA encoding the required functions is specifically synthesized. The relevant genes in this operon have been cloned and sequenced and their products have been well characterized. With the advent of genetic engineering the most elegant experiments have been carried out using synthetic genes which have been specifically constructed to study the individual components of the operon. While numerous questions remain, insights into the controls of development of organisms and differentiation of cells are coming at a dazzling speed. This area was just a short while ago one of life's most well-guarded secrets. Undoubtedly numerous surprises lie ahead. Indeed, a most recent surprise has been the discovery that many of the genes in higher organisms are not contiguous. This was first noticed in 1977 [9-12] when different species of the mRNA coding the same gene were observed. It seemed that the mRNA was being processed and that non-coding pieces were being removed from the midst of coding sequences. This process, which continues until an intact coding sequence remains, is illustrated in Figure 11. Not every

PATHWAY OF mRNA MATURATION

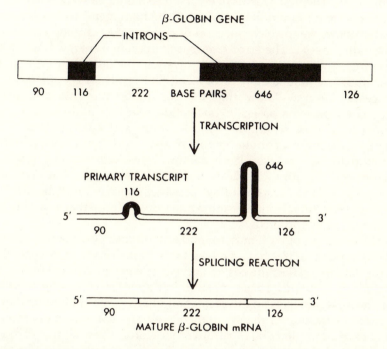

Figure 11 Many genes in eukaryotic cells are disrupted by the presence of non-coding sequences called introns. The mRNA produced by such genes must be "processed" into their mature form so that the coding region is continuous.

gene has such interruptions, which are called introns, and while the human β-globin gene contains only two, the gene for collagen, a protein about 900 amino acids long, possesses nearly 40 introns! The mechanisms by which processing occurs is beginning to be understood, but the reasons why genes have such non-coding inserts remains largely a question of conjecture.

Conservation of Genetic Material

Mutation is the driving force of evolution. At this time we are exploring the sources of mutation and investigating the cellular defences against genetic change. The discovery of the double helix was followed almost immediately by the recognition that errors could occur during DNA replication. The first sources of errors to be identified were the natural tautamers and isomers of DNA which could hydrogen-bond incorrectly and result in the misincorporation of an incorrect, uncomplementary base into the DNA. The error-rates based upon free energy differences of correct and incorrect nucleotides were in the order of 1 per 100. It was therefore immediately recognized that synthesis is not a passive process. We now know that the replication enzymes and their associated proteins assist in nucleotide selection and enhance the accuracy of DNA replication. The isolation and characterization of highly inaccurate or mutator polymerases and extremely accurate, antimutator polymerases clearly demonstrates a role for these enzymes in the accuracy of DNA replication. Indeed, polymerase functions are known to include "proof-reading" activities which monitor the correctness of the last nucleotide inserted and remove those that appear to be incorrectly incorporated.

The existence of both mutator and antimutator polymerases illustrates that polymerases have evolved to an optimal error-rate. Generally this is taken to reflect the cost of surveillance versus the cost of errors. The recent sequencing of several hundred spontaneous mutants in our lab, however, reveals an interesting alternative. We are finding that a significant fraction of the spontaneous mutation arises as the consequence of attempted dna repair. In other words, the evolution of more accurate polymerases would not contribute significantly to a reduction in the overall mutation frequency of an organism. Error-rates measured *in vivo* are extremely low, on the order of one error per billion base pairs replicated. Curiously, the error-rate of the most accurate DNA polymerases measured *in vitro* are 100-fold greater.

This observation suggested that other mechanisms were responsible for the greater fidelity (to an extra three orders of magnitude) seen *in vivo*. One mechanism available to the cell is what we have called mismatch repair. This is a process by which incorrectly paired nucleotides are recognized and removed from the DNA. The model involves postulating the existence of enzymes capable of recognizing misincorporated bases and directing their removal. Our model proved correct and we were able to demonstrate this repair process [13]. When we isolated mutants defective in mismatch repair we were delighted to discover that they were what we call "mutators" and had mutation frequencies 100 to 1000 times above normal. We therefore believe mismatch repair to be an important contributor to the overall accuracy with which DNA is replicated.

Numerous repair processes and error-avoidance mechanisms have been identified. These include excision repair, in which bent and battered nucleotides are physically removed from DNA and replaced. A class of enzymes called glycosylases removes tattered bases. The classic example is the enzyme Uracil-DNA-glycosylase, which removes uracil from DNA. This is an important mechanism of error-avoidance; even at 37°, cytosine spontaneously deaminates to uracil at measurable frequencies. Guanine is paired with cytosine in DNA; however, uracil behaves like thymine and, in the next round of replication, would pair with adenine. Thus, what was a G:C base pair would become an A:T base pair. In other words, the deamination of cytosine in DNA results in a mutation. The removal of uracil from DNA is the cell's way of preventing mutation. This is not a trivial repair process. Every twenty-four hours, more than 3,000 spontaneous deaminations occur in every cell in our body. Evolution has provided us with an effective defence against this source of genetic destabilization. Indeed, the DNA is subject to insult from endogenous and exogenous sources too numerous to catalogue. Curiously, DNA repair was perhaps the only aspect of DNA metabolism not predicted by the great sages of molecular biology.

Antibody Diversity: The Exception That Proves the Rule

Mammals produce millions of different antibodies over the course of a lifetime. We know that the genes producing the antibodies diversify during the development of the organisms. This diversity is obtained by a variety of DNA-metabolic processes. These include (1) recombination of gene subunits permitting great variation; (2) variation in the selection of the joining fragment that brings together the variable portion of the gene (accounting for the greatest portion of diversity) and the constant region (required for the functioning of antibodies); (3) the wealth of variation available from the diversity of variable genes; and (4) variations in mRNA processing which produces different mature mRNAs. The pertinent point here is that the variable gene arises by spontaneous somatic mutation. This is an extraordinary mechanism for generating diversity on command. Normally, the cell makes as great an effort as possible to keep mutation levels as low as possible. In this instance, however, the organism is able to remove a small, specific fraction of the DNA from the normally stringent requirements for fidelity and conservation of the genetic material. This demonstrates beyond a doubt that the fidelity of DNA replication can be regulated by the cell.

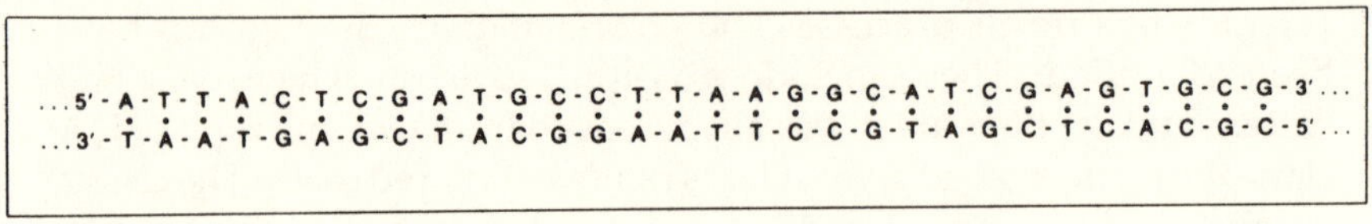

...5'-A-T-T-A-C-T-C-G-A-T-G-C-C-T-T-A-A-G-G-C-A-T-C-G-A-G-T-G-C-G-3'...
...3'-T-A-A-T-G-A-G-C-T-A-C-G-G-A-A-T-T-C-C-G-T-A-G-C-T-C-A-C-G-C-5'...

Figure 12 The DNA is diagrammed here in its ordinary double-stranded configuration. The bases are appropriately paired and the hydrogen-bonding occurs between two complementary strands of DNA.

Sequence Directed Mutational Events

One of the most exciting studies in my laboratory has been work demonstrating that DNA sequences can directly influence their own mutational fate [14–16]. Such data derive from our ability to clone and sequence mutants at a rapid rate. This mechanism is shown in deletion mutations we have sequenced. First, it must be emphasized that DNA need not always be the linear double-stranded molecule exemplified in Figure 12. Under some circumstances palindromic DNA sequences permit the DNA to assume other topological configurations. Palindromes are sets of sequences which are complementary, but in reverse order. They permit the formation of aberrant structures such as those illustrated in Figure 13, in which the nucleotides are properly hydrogen-bonded but misaligned. In the case of palindromes, the aberration is that the hydrogen bonds are not between different strands of DNA but within the same strand. Other misalignments are also possible. For example, misalignments can occur between repeated sequences. In this case, one copy of a repeated sequence is misaligned onto the complement of the second repeat. As you will see, such misalignments represent a challenge to the integrity of DNA and are capable of directing specific classes of mutational alterations.

Among a total of twenty-six deletions lying within the *lacI* gene we note that twenty-two have the following common. misalignment configuration. The deletions are delineated by repeated sequences, and potential structural intermediates capable of accounting for the mutation can be drawn. However, in each of the twenty-two cases, this misaligned structure is further stabilized by the presence of palindromic sequences. A typical example is shown in Figure 14. The fact that both repeated and palindromic sequences are at the termini of so many spontaneous deletions argues strongly for their involvement in mutational events. This is an important consideration because it shows that DNA sequences can direct their mutational fate.

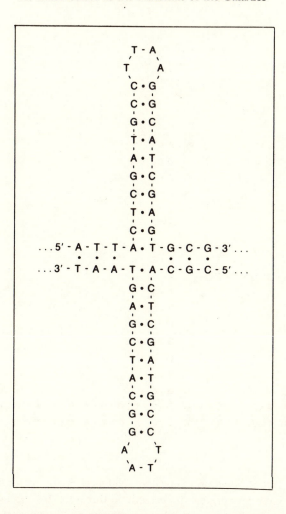

Figure 13 The sequence presented in Figure 12 is palindromic (or self-complementary) and can therefore take on an alternative structure known as a hairpin. Here the hydrogen bonds form between base pairs in the same strand of DNA.

Not only deletions have their origin in hairpin structures. We have also recovered frameshift mutations whose origin is best explained as the consequence of attempted repair events involving hairpin structure [15–17]. Such an example is shown in Figure 15. In this case mispaired bases in a misaligned structure may provide substrates for the same pathway of repair as we observe during mismatch repair. We have

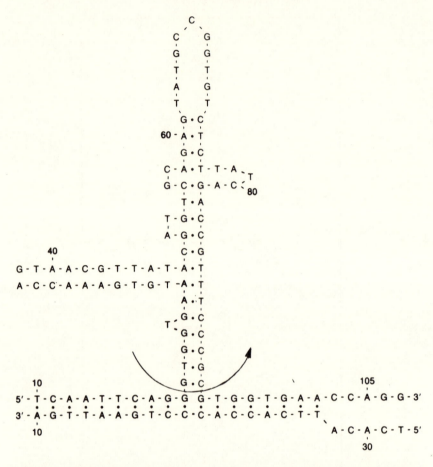

Figure 14 Palindromes and repeated sequences make DNA misalignments possible. Such misalignments present a threat to the accuracy of DNA metabolism. The figure shown here demonstrates the suspected structural intermediate of a real deletion mutation isolated in the *lacI* gene of *Escherichia coli*. In this structure both palindromic and repeated sequences contribute to the misalignment. The palindrome consists of those bases making up the stem of the hairpin structure that are hydrogen-bonded. The repeated sequences are G-T-G-G-T-G-A-A. This sequence is found once in the 5′-side of the palindrome and again on the 3′-side at its base. There the repeated sequence is actually hydrogen-bonded to the complement of the first copy of the repeat. The frequency with which both palindromes and repeated sequences both contribute to the stability of the structural intermediates of deletion indicates the importance of their role in the formation of deletion mutants.

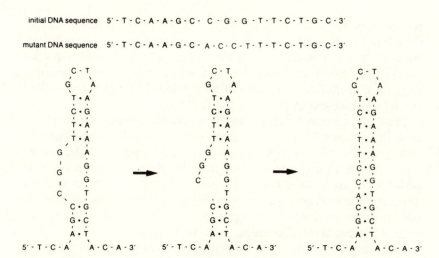

initial DNA sequence 5' - T - C - A - A - G - C - C - G - G - T - T - C - T - G - C - 3'

mutant DNA sequence 5' - T - C - A - A - G - C - A - C - C - T - T - T - C - T - G - C - 3'

Figure 15 An imperfect palindrome can generate a misaligned DNA structure that provides a template for frameshift mutation. The removal of the imperfect portion of the palindrome may permit resynthesis using the opposite portion of the hairpin stem as a template. In the case shown here a C-G-G sequence is lost and replaced by A-C-C-T. This model explains the occurrence of complex mutational events involving simultaneously both base substitution and frameshift mutations.

further demonstrated that duplications and insertion events can have their origin in misaligned structures. We have coined the term "sequence directed," since DNA sequences potentiate the misaligned structures that predict the mutational events.

We have also accumulated some evidence that misalignment mutagenesis may be of importance in mammalian cells as well. In collaboration with Dr G.B. Golding at York University, we aligned a collection of fifteen human leukocyte interferon genes and inferred a phylogenetic history of these genes. As we have already seen from the examples presented, a diagnostic feature of sequence-directed mutational events is the prediction of closely linked concurrent mutations. Several multiple mutations are inferred to have occurred during the evolution of the interferon genes and for these events either palindromic or directly repeated sequences are available which could have templated their occurrence. This is illustrated with two examples. From this study we concluded that the DNA sequence has the potential to account for most if not all mutations by a mistemplating mechanism. In particular, the results of our study indicate that complex alterations to sequence likely reflect sequence-directed mutational events. These mechanisms suggest that mutation would not be random but rather dependent upon local DNA sequence.

The tools of modern biology enable us to design an experiment to test such hypotheses. I can safely argue that these are the most elegant of "Designer Genes." By a process known as site-specific mutagenesis, and taking into account the fact that the code is degenerate, it is possible for us to alter DNA sequences at specific positions without altering the amino acid sequence of the protein. As a consequence, we have begun to evaluate the potential contribution of different components of structural misalignments in mutagenesis.

In addition, the advent of biotechnology now permits the study of mammalian genes and the direct analysis of mammalian mutations. We have begun the analysis of the specificity of spontaneous mutations by directly cloning and sequencing *de novo* mutations in the *aprt* gene in Chinese hamster cells. We have now established the DNA sequence of the normal, active *aprt* gene. We have characterized the control regions, the coding regions, the four non-coding regions (or introns) and several hundred base pairs up and down from the gene. We have also cloned some fifty spontaneous mutants and have established the mutant sequence for many of these. As data accumulates we will gain further insight into the mechanisms of mutation operating in mammalian cells.

We cannot help but be excited by the possibilities. Not only will the analysis of such collections of mutants enable us to learn more about the sources of mutation and the cell's ability to carry out DNA repair, but also mutations in non-coding regions will help us to unravel the mysteries of gene structure and gene expression. The ability to examine genes, the building blocks of life, can be considered no less a wonder than examining the secrets of the heavens or the inner workings of the atom.

Where Are We Headed?

Molecular biology has contributed greatly to our understanding of the process of evolution. The power of inference in discerning ancestral sequences will continue to amaze as the number of sequences available increases. Perhaps of less consequence, but certainly of dramatic proportions, is the ability of scientists to clone and sequence the DNA of long-extinct animals. Nonetheless, the remarkable rate with which this science is progressing feeds my anticipation. We will soon have more information of the nature of life than we could have thought possible even ten years ago. Indeed, the gene has become more tangible to us than the material of the galaxies, perhaps even more so than oxygen was to Priestly two centuries ago. Frankly, I feel like a tourist in a European city who eagerly reads the menus in the windows of restaurants not yet open.

Reflect upon it for a moment. We are at the edge of a fundamental revolution. The species barriers have been crossed. Mice carrying human genes, for example, the human growth hormone gene, have been constructed [18]. New life forms have been patented. Bacteria have been created that produce human insulin. Site-specific muta-genesis is a standard technology and gene therapy is as plausible as blood transfusions were just a few decades ago. We are assembling the tools with which we can alter the flow of evolution. The study of mutation can, within our working lives, change from a passive to an active enterprise. Indeed, mutation, a word meaning change, may regain the power of the Latin equivalent, *alchemia*. Our capacity to change life forms rivals in significance our power to create the atomic bomb. Humankind, if it chooses to have a future, now has the opportunity, and obviously the responsibility, of shaping that future. Indeed, as it is written:

And God blessed Noah and his sons and said to them, "Be fruitful and multiply, and fill the Earth. The fear of you and the dread of you shall be on every beast of the Earth and bird of the air, upon everything that creeps on the ground and all the fish of the sea; into your hand they are delivered. Every moving thing that lives shall be food for you; and as I gave you green plants, I give you everything." (Genesis)

REFERENCES

1 T.H. Morgan. 1910. "Sex-Limited Inheritance in Drosophila." *Science* 32:120–2.
2 H.J. Muller. 1927. "Artificial Transmutation of the Gene." *Science* 46:84–7.
3 L.J. Stadler. 1928. "Mutations in Barley Induced by X-rays and Radium." *Science* 68:186–7.

4 O.T. Avery, C.M. MacLeod, and M. McCarty. 1944. "Studies on the Chemical Nature of the Substance Inducing Transformation of Pneumococcal Types." *J. Exp. Med.* 79:137–58.

5 A.D. Hershey and M. Chase. 1952. "Independent Functions of Viral Protein and Nucleic Acid in Growth of Bacteriophage." *J. Gen. Physiol.* 36:39–56.

6 J.D. Watson and F.H.C. Crick. 1953. "Molecular Structure of Nucleic Acid: A Structure for Deoxyribose Nucleic Acid." *Nature* 171:737–8.

7 J.D. Watson and F.H.C. Crick. 1953. "The Structure of DNA." *Cold Spring Harbor Symp. Quant. Biol.* 18:123–31.

8 C. Yanofsky, B.C. Carlton, R.J. Guest, D.R. Helsinki, and U. Henning. 1964. "On the Colinearity of Gene Structure and Protein Structure." *Proc. Nat. Acad. Sci. U.S.A.* 51:266–72.

9 S.M. Berget, A.J. Berk, T. Harrison, and P.A. Sharp. 1978. "Spliced Segments at the 5' Termini of Adenovirus-2 Late mRNA: A Role for Heterogeneous Nuclear RNA in Mammalian Cells. *Cold Spring Harbor Symp. Quant. Biol.* 42:523–9.

10 T.R. Broker, L.T. Chow, A.R. Dunn, R.E. Gelinas, J.A. Hassel, D.F. Klessig, J.B. Lewis, R.J. Roberts, and B.S. Zain. 1978. "Adenovirus-2 Messengers – An Example of Baroque Molecular Architecture." *Cold Spring Harbor Symp. Quant. Biol.* 42:531–53.

11 R. Breathnach, J.L. Mandel, and P. Chambon. 1977. "Ovalbumin Gene is Split in Chicken DNA." *Nature* 270:314–19.

12 A.J. Jeffreys and R.A. Flavell. 1977. "The Rabbit β-globin Gene Contains a Large Insert in the Coding Sequence." *Cell* 12:1097–1108.

13 B.W. Glickman and M. Radman. 1980. "Escherichia Coli Mutator Mutants Deficient in Methylation-instructed DNA Mismatch Correction." *Proc. Natl. Acad. Sci. U.S.A.* 77:1063–7.

14 B.W. Glickman and L.S. Ripley. 1983. "Structural Intermediates of Deletion Mutagenesis: A Role for Palindromic DNA." *Proc. Natl. Acad. Sci. U.S.A.* 81:512–16.

15 W.J. Drake, B.W. Glickman, and L.S. Ripley. 1983. "Updating the Theory of Mutation." *American Scientist* Nov.-Dec.: 621–30.

16 L.S. Ripley and B.W. Glickman. 1983. "The Unique Self-complimentary of Palindromic Sequences Provides DNA Structural Intermediates for Mutation." *Cold Spring Harbor Symp. Quant. Biol.* 47:851–61.

17 L.S. Ripley. 1982. "Model for the Participation of Quasi-palindromic DNA Sequences in Frameshift Mutation." *Proc. Natl. Acad. Sci. U.S.A.* 79:4128–32.

18 R.D. Palmiter, G. Norstedt, R.E. Gelinas, R.E. Hammer, and R.L. Brinster. 1983. "Metallothionein-Human GH Fusion Genes Stimulate Growth of Mice." *Science* 222:809–14.

BIBLIOGRAPHY

Freifelder, D.A. 1983. *Molecular Biology: A Comprehensive Introduction to Prokaryotes and Eukaryotes*. Portolo Valley, Calif.: Jones and Bartlett.

Hanawalt, P.C. 1973. *The Chemical Basis of Life: An Introduction to Molecular and Cell Biology – Readings from "Scientific American."* San Francisco: W.H. Freeman.

Judson, H.F. 1979. *The Eight Days of Creation*. New York: Simon & Schuster.

Kornberg, A. 1980. *DNA Replication*. San Francisco: W.H. Freeman.

Watson, J.D. 1968. *The Double Helix*. New York: Atheneum.

Watson, J.D. 1976. *Molecular Biology of the Gene*. 3rd ed. Menlo Park, Calif.: Benjamin/Cummings.

W. FORD DOOLITTLE

From Selfish DNA to Gaia: One Molecular Biologist's View of the Evolutionary Process

1 Introduction

1.1 *The view from molecular biology.* To begin, a confession. I was trained as an experimental molecular biologist during the 1960s. It was understood that I was to believe that nothing really important happened in biology until 1953, when Watson and Crick came up with a structure for DNA; that this discovery made flesh the words of Darwin, which were never to be doubted (nor, except out of curiosity, read); that an interest in the history, or worse the philosophical roots, of the discipline was a symptom of intellectual weakness; that the whole was *never* more than the sum of the parts; and that grown men never cry. To be sure, molecular biology began with somewhat loftier intents – the hope that an understanding of heredity would require at least some new general laws of physics, for instance. However, most molecular biologists were too young to read when Schrödinger wrote *What is Life?* (1945), and today we are mostly preoccupied with the molecular nuts and bolts of the biological machinery and generally unconcerned even with issues in the larger context of evolutionary biology within which these data are best evaluated.

My own contact with anything beyond these narrow prescriptions came through a consideration of how natural selection might act upon the molecular structure of DNA. This led to more general interest in questions of the organizational levels (gene, individual, species) at which selection can take place, and of interactions between levels –

questions in which real evolutionary biologists and some philosophers have taken interest. In all this I am an amateur: the best I can do is present those metaphysical attitudes which the practice of molecular biology *can* engender in the unwary practitioner, and to outline a way of thinking which might get us out of the resultant existential pit. One of molecular biology's genuine heroes, Jacques Monod (1971), after 20–30 years of delineating the intricacies of the coupling between genotype and phenotype, ultimately came to believe that "man at last knows that he is alone in the unfeeling immensity of the universe . . . Neither his destiny nor his duty have been written down." I for one am uncomfortable with that view.

1.2 *Faith*. Faith, it seems to me, is something one either has or does not have, and it derives from experience, not from the exercise of reason. Thus, I guess that I donot believe that science will reveal any gaps through which God will appear to convince the non-believer. Like many scientists, however, I often have *too much* faith in the explanatory power of the scientific method. I forget that the method is by definition reductionist and materialist and that I have no right and no reason to expect to find myself in any situation other than "alone in the unfeeling immensity of the universe" if I adopt it as personal metaphysics. I derive considerable comfort from two recent popular books, Paul Davies' *God and the New Physics* (1984) and Hoimar von Ditfurth's *The Origins of Life* (1982). Neither writer attempts to *prove* that the universe is other than unfeeling and neither proposes any alteration of the *scientific* interpretation of scientific data, but both have changed the way I *feel* about the universe and the data – and perhabs this is all I can expect. I find it helpful to be reminded that my mind is the product of still incomplete evolutionary process, and thus that I cannot expect from it a clear and complete picture of reality; that the naive realism of my outlook as a molecular biologist cannot cope at all with quantum uncertainty or wave-particle duality; and that I am thus free to construct my own mythology. As Alex Comfort points out in *Reality and Empathy* (1984), another good, recent, popular book, "there is no rational world model which would be wish-fulfilling, but some models pinpoint or irritate our existential malaise more than others, and *some* world model is necessary in order to live."

The situation is clearly better in physics, where there seem to be some truly counter-intuitive imponderables, than in contemporary biology. Here, at a level above and perhaps unaffected by quantum uncertainties, but still below and thus unperturbed by the centuries-old mind-body paradox, there seems to be nothing that we cannot, at least in principle, explain – no "gaps for God." It is of course logically permissible to view the data of modern biology in the soothingly holistic way that writers such as Lewis Thomas do, but such views simply do not, I submit, emerge in any natural way from the data or the methodology themselves. I can think of just a few intellectually honest ways out. One is to reduce biology to physics and let a sense of wonder and mystery creep back in, as it were, from below. Another is to see all of evolution, both physical and biological, as an ongoing act of creation, as Teilhard de Chardin and more recently, and to my mind more convincingly, von Ditfurth have done. A third and more interesting way might emerge if Darwinian theory could be incorporated within a larger and less confining way of thinking about evolution and selection, in the same way as Newtonian physics has been incorporated within a larger theory of quantum and relativistic physics, and with the same kinds of potential spiritual advantage. I have no way of knowing whether this is possible. If there is any hope at all, I suspect it will come through thinking about selection as it affects supra-individual biological units, that is through coming to terms with Lovelock's Gaia. So, after a lengthy discussion of the nature of selection and the current tendency to consider selection at increasingly lower levels of biological organization, I shall try to see if there is anything more "holistic" one might do.

2 Views on Selection

2.1 *Selection in general.* Evolution can be seen as the result of two processes, the generation of variants and the differential reproduction of those variants. Together these processes are in a real sense creative, and result in the appearance of design without a designer. They are potentially capable of accounting for any number of watches found on any number of heaths and this is one reason why Darwinism was initially, and often still is, seen as inimical to faith. Dr Haynes discusses the creative role of the first process, the generation of variants by events which are, at least at the level of the organism, chance events. I will deal more with the second – the differential reproduction of variants – which in fact is the principle of natural selection.

Differential reproduction (evolution by natural selection) can result from either differential viability or differential fertility, but we can treat the effects of these, and of any heritable traits related to them, together and call them in sum reproduction – if we mean by that the overall act of replication (the production of new individuals, rather than the more biologically specific act of procreation, by whatever method).

Selection results automatically in the *relative* increase, within successive populations, of types that are more successful at surviving or are more fertile, and the relative decrease of types that are less successful or less fertile. If population size does not increase indefinitely, if for instance it is constant from generation to generation, then of course natural selection will effect the *absolute* increase of fitter types, and the absolute decrease, ultimately to extinction, of less fit types. It is this imposition of the Malthusian parameter which makes evolution into a "struggle for existence" in which only the fit survive. The apparent ruthlessness of this is of course still another reason why Darwin's views encountered religious opposition – there seems no place for intraspecific sharing and caring in such a Malthusian world.

As almost everyone who writes about Darwin is quick to point out, vicious paw-to-paw combat is seldom at issue here. Darwin (1859) wrote: "I use the term Struggle for Existence in a large and metaphorical sense, including dependence of one being on another, and including (which is more important) not only the life of the individual, but success in leaving progeny. Two canine animals in a time of dearth, may be truly said to struggle with each other which shall get food and live. But a plant on the edge of a desert is said to struggle for life against drought." Elliot Sober, in his masterful work *The Nature of Selection* (1984), likens the two kinds of struggle to the games of tennis and golf. In the former, the winner gains points at the expense of the loser. In the latter, both players score against an external standard, but still there are winners and losers.

The principle of selection is truly general; it is law of nature. Any entities (organisms, business firms, religious sects, abstract ideas) that in some way replicate themselves (in nature, in the market place, in the hearts and minds of men and women) producing replicas which are however slightly different from themselves (by mutation, alteration in management, revision of dogma, intellectual elaboration) will *evolve*, provided of course that these differences are passed on to replicas of the replicas. They will evolve by *natural selection* if these differences in some way affect success within the milieu in which replication occurs (if, for instance, in the case of abstract ideas, differences in formulation affect comprehensibility and thus acceptability).

As Sober clearly shows, Darwin's theory of evolution actually makes two claims. First, that any collection of entities which shows the above characteristics (which exhibits *heritable variation in fitness*) will of necessity evolve by natural selection. Second, that living organisms show and have shown, throughout the history of life on earth,

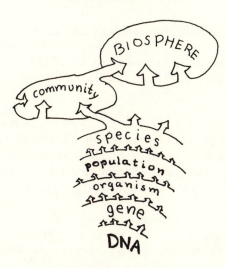

Figure 1 Levels of biological organization and selection

heritable variation in fitness, and this has been the primary force (not necessarily the only force) responsible for the diversity of species we now see.

This all seems so simple that sometimes the whole concept just slips away – seeming as true but also as empty as our knowledge that 2 + 2 = 4, or that effects have causes – and some people fall into the heresy of calling the principle tautological. Philosophers have dealt with this objection at length, and overcome it (Mills and Beatty 1979; Williams 1973). I am persuaded by their arguments, although I do not really understand them. Sober (1984) entertains the possibility that the concept of natural selection is in fact *a priori*, not an empirical hypothesis at all. I find this fascinating.

2.2 *Levels of organization and selection.* The biological world is hier-archically organized (Fig. 1) and entities showing heritable variation in fitness can be defined, with varying degrees of coherence, at all levels. Here I distinguish DNA, genes, organisms, populations, species, com-munities, and biospheres as levels. Clearly, however, one could cut things thicker or thinner, and the boundaries *are* fuzzy. Entities at each level interact with each other (horizontally) through mechanisms and by rules which may be related only by analogy to mechanisms and rules found at other levels. What happens at each level affects (vertically) what happens at other levels, through still different kinds of mechanisms and rules. The reductionist approach compels us to view the chain of causation here as going from lower to higher. Almost always we think of events at one level generating the random variants amongst which selection selects at the next level higher.

2.3 *The modern synthesis and the role of molecular biology.* The most obvious kinds of biological entities with heritable variation in fitness are individual whole organisms, and it was chance variation amongst organisms, with selection thus determining the composition of populations and through that the characters of species, to which Darwin addressed himself.

The theory was incomplete, however, without subsidiary theories about mechanisms of heredity (Mendelism) and the nature of genes (the central dogma of molecular biology), and without an understanding of the behaviour of genes in natural populations, which required both theoretical and experimental developments in population genetics. All of them went to make up the "Modern Synthesis" (Mayr 1982). Molecular biology's particular role in this synthesis was to explain the coupling between *genotype* and *phenotype* (Fig. 2).

An organism's genotype is simply the sum of all the genetic information it has, the aggregate of all its genes. Its phenotype is really just itself: its structure, its physiology, its biochemistry, its mode of reproduction, its patterns of parental care, its metaphysical concerns. Genotype does not completely determine phenotype because some of these things are influenced by environmental factors such as diet, the availability of mates and philosophical discussions. Furthermore, we do not really understand how the interactions of multiple genes are combined so as to produce, through development, whole animals and plants. But we do know that DNA $\rightarrow$ RNA $\rightarrow$ protein, and also that we can *in principle* account for all of an organism's physical form and physiological function, and perhaps much of its behaviour, in terms of amounts, structures, and chemical functions of proteins. Thus we can also in principle understand all of evolution, as follows. Random variations in DNA sequences (chance mutations in genes) produce random changes in protein structure and function, which produce random changes in overall phenotypes of organisms. Selection picks organisms of fitter phenotype, and their favoured reproduction differentially propagates the initially variant DNA sequence (mutant gene). We also know that, although we can imagine mechanisms by which phenotype could feed directly back to genotype (protein $\rightarrow$ RNA

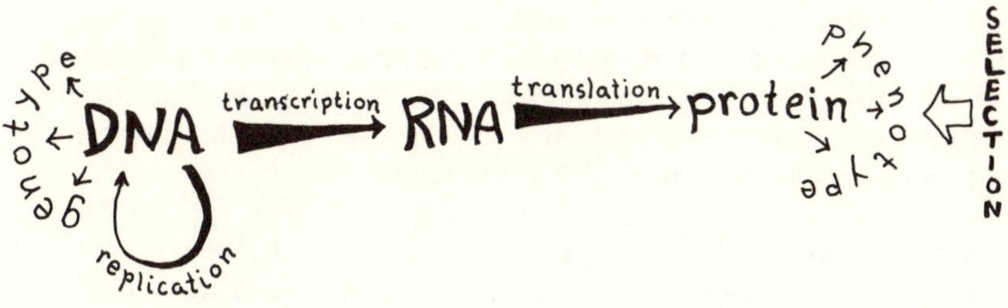

Figure 2 The central dogma of molecular biology, elaborated

→ DNA), such mechanisms are not actually in place in any known organism. Thus Lamarck was wrong, and it is probably never possible for an organism to pass on its mortal experience or aspirations to its progeny, by genetic means.

There is no logical reason why all of this should be more discouraging than the knowlege that the earth goes around the sun, but somehow it is. Having taken the watch apart, we see that there is nothing all that special about it. It would take no vital essence to put the watch back together, only time and money. There are no gaps for God here, and it is possible to feel alone and afraid.

2.4 *Genic versus genotypic selection.* There is, however, a certain wholeness in the concept of the organism and a certain simple elegance in the coupling of genotype to phenotype. Lewontin (1974) and Mayr (1982), for instance, have emphasized the cohesion of the genotype, the dependence of the phenotypic expression of any one gene upon the expression of many others, and thus the irreducibility of the phenomenon of biological evolution below the level of the whole organism. This was of course Darwin's position.

But Darwin did not know about genes, and it is perhaps most common now to see selection as acting on individual genes, not whole organisms, and the level of the gene as the level at which the important events of evolution really occur. This attitude is a natural outgrowth of the theoretical structure of mathematical population genetics, which studies changes in gene frequencies and the (hypothetical) selection values of single alleles (mutants) of single genes. In fact it is obliged to do this, for reasons of methodological manageability, but it is very tempting to take this reductionist research strategy – Mayr calls it "bean bag genetics" – as an ontological conclusion. The writings of

Richard Dawkins show what happens when one does this. "We are survival machines – robot vehicles blindly programmed to preserve the selfish molecules known as genes ... they are in you and me; they created us, body and mind; and their preservation is the ultimate rationale for our existence" (Dawkins 1976).

2.5 *Selfish DNA*. Although this gene-centred view reduces the level at which we see biologically significant selection as occurring, Dawkins is simply constructing a grisly metaphor out of the research program of population genetics. In this model, even selfish genes must usually look after, must care for and maintain, the survival machines that reproduce them; and usually they must cooperate with other selfish genes in doing so. That is, it is still the case (with minor exceptions) that for a gene to do better in the evolutionary struggle than competing genes, the organism it inhabits must do better – genotype must still benevolently determine phenotype.

This is not so if we reduce our thinking about selection one level further. In doing this, Carmen Sapienza and I (1980), and Leslie Orgel and Francis Crick (1980), came up with something called the "selfish DNA" hypothesis. To see what we were driving at, it is only necessary to realize that the inside of the nucleus of a cell is an environment for the DNA which replicates there, just as ecosystems are environments for the organisms which replicate there. Selection can occur within nuclei just as it occurs among the bushes and in the swamps and as it did in the primeval ocean. In this (nuclear) environment there are entities which show *heritable variation in fitness* and which will evolve by natural selection. These entities are pieces of DNA, and it does not matter for our argument that they may be all strung together as parts of chromosomes. Some of these pieces are genes: they find expression in organismal phenotype (DNA → RNA → protein) and have to account for themselves to natural selection operating at the level of organismal phenotype. But most of these pieces are not genes; 90 to 99 per cent of the DNA of most organisms is not expressed in phenotype. Unexpressed DNA is replicated along with the genes, by the same mechanisms, and as part of the same chromosomes, and is subject to variation in sequence (mutation), just like the genes.

The unexpressed pieces thus show *heritable variation*. They would show heritable variation *in fitness* – that is, they would evolve by natural selection – if there were some way other than phenotypic expression by which they could influence their own likelihood of survival, by which they could be defined as more or less fit. The biggest threat to survival is deletion, random excision from the chromosomes and thus exclusion from access to the vital machinery of DNA replication. (This happens often in evolution and the organism as a whole presumably benefits – becomes more efficient – as a result of such molecular pruning.)

There *is* a way for a piece of DNA to evade deletion, and this is to over-replicate, to deposit extra copies of itself in other locations in the chromosome complement. If these extra copies can in turn make extra copies, then deletion could be surely outrun and we would have what look for all the world like genetic parasites, parasites that need have no beneficial effect on organismal phenotype (and that could in fact have deleterious effects).

Mechanisms for over-replication are known; transposition (Fig. 3) is the most easily described. All that is required further, for evolution to proceed by natural selection, is for the exact structure, the DNA sequence of the pieces of DNA which are to be transposed, to determine the likelihood that they *will* transpose – so that heritable variations can affect fitness. And this too is found. Organisms of all kinds (bacteria, animals, and plants) contain transposable elements which over-replicate sufficiently often to outrun deletion and whose ability to transpose depends on sequence – often in very elaborate ways (Shapiro 1983). Many transposable elements carry DNA sequences that determine the structures of enzymes which catalyse their own transposition. (These enzyme-determining sequences are of course "genes," but in a special way, since the "phenotypes" they affect are those of the transposable elements themselves, not of the organisms whose genomes they parasitize.)

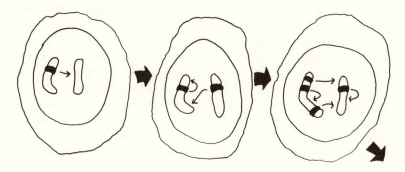

Figure 3 Schematic representation of transposition, through successive cell (or organismal) generations

Not all of the 90 to 99 per cent of an organism's DNA that is not phenotypically expressed is made up of such transposable elements – of what we call *selfish* DNA. But a surprising amount (often more than the amount needed for real genes) is. Much of the rest of the DNA can be accounted for by other mechanisms of over-replication. These mechanisms, too, seem to have nothing to do with the maintenance of organisms but everything to do with the maintenance of DNA.

This carries us one step further than Dawkins. Selfish genes are obliged at least to ensure that we, their survival machines, function well and give the appearance of design. Selfish DNA does not bother with this nicety. If we choose to construct an ontological metaphor out of this, it will not be very cheery. What evolution is really about is DNA and the replication and transposition of DNA. Phenotypes across the entire spectrum from the ability to utilize glucose to the ability to wonder about God are, in a way, very nearly dispensable epiphenomena.

Research programs are of course not metaphysics, and we can retain faith in spite of the harsh facts of molecular biology. All I hope to have accomplished here was to show that these facts, *interpreted within the reductionist program we have adopted*, are pretty harsh, harsher even than Darwin's "struggle for existence." The only natural theology that emerges naturally from them is what Tillich (1952) has called "the courage of despair."

3 Other Programs: The Possibility of Gaia

3.1 *How we think.* The extent to which my own way of thinking has been influenced by the methodology of my discipline was made apparent to me in a recent conversation with Anthony Barton, one of Lovelock's strongest supporters. His perceptions, his notions of reality, seem entirely different from mine, and yet there is no disagreement between us about ascertainable facts, only about how they are to be interpreted. His perception is holistic, mine of course reductionist. I am quite confident that mine has more explanatory power and would suggest more hypothesis-confirming experiments. But for just an instant I found myself thinking like Barton (or, rather, thinking in a way which would lead me to talk like Barton) and I cannot for the life of me say that this way of thinking felt in any way less rational or less "right" than my usual way.

3.2 *The trouble with Gaia.* Lovelock develops the notion of Gaia from a consideration of the remarkable (as compared to expectation or as compared to lifeless planets) constancy of environmental conditions on earth, since near the time of the origin of life (about 3.5 billion years ago). There are many global physical and chemical parameters which should have varied outside ranges hospitable to life as we know or could imagine it which, demonstrably, have not so varied. Clearly the presence of life had a lot (maybe everything) to do with this maintenance of constancy (Lovelock 1979).

Lovelock invokes in Gaia, "a complex entity involving the Earth's biosphere, atmosphere, oceans, and soil; the totality constituting a feedback or cybernetic system which seeks an optimal physical and chemical environment for life on this planet." The Gaia hypothesis "postulates that the physical and chemical conditions of the surface of the Earth, of the atmosphere, and of the oceans has been and is actively made fit and comfortable by life itself." In short, Lovelock sees Gaia as an homeostatic system maintained "*by and for*" the biosphere (Lovelock 1979). I have no trouble with the *by*, but as a reductionist and a Darwinist, I have a lot of trouble with the *for*. This is not because of its teleological aspect. When an evolutionary biologist says that an organism does something *for* a purpose, he really means that selection has favoured, in populations of the organism, genes that (by chance) accomplish that "purpose." Everyone understands this, and a reductionist would say, in this case, that natural selection has favoured genes which through their expression contribute to global homeostasis; I assume that that is what Lovelock means by *for*, also. The problem with *for*, rather, is this: because the feedback loops Lovelock envisages are so complex and so global, and especially because they affect events over such long time scales, it is impossibly difficult to see how selection operating at *any* level could produce them.

Let us take as an example the scenario for maintenance of the oxygen content of the atmosphere constructed by Watson, Lovelock, and Margulis (1978). Oxygen is produced by photosynthesizers, and it is kept in check by combination with methane, the product of methanogenic bacteria. Watson et al. describe a mechanism by which elevated oxygen concentrations indirectly promote the growth of methanogens, and thus the production of oxygen-reducing methane. The scenario may well be correct, but there is nothing really in it, for instance, for the methanogens. If they stop producing methane, or stop responding to oxygen concentration, as they would if selection at the level of individuals promoted the spread of non-producing or non-responsive mutants, nothing adverse would happen for a very long time. I do not know enough about global cycles to say how long, but surely millions or billions of methanogen generation-times. By this time, no producing or responding bacteria would be left. Organisms do not evolve, by natural selection, traits which are only beneficial to their descendants millions of generations down the line, nor do they maintain traits simply because, millions of generations later, they will be sorry they had not. Evolution lacks foresight. There is no effective selection pressure operating on methanogens to contribute to the homeostasis as such. Of course methanogens may (and do) produce methane for selfish reasons of their own, but there is again no selection for the cybernetic loop; if it exists it is accidental.

The same argument can be applied to any loop or collection of loops. No standard reading of the Darwinian paradigm will allow there to be genes in organisms that owe their existence to selection *for* global homeostasis. Although Lovelock's thinking on this has clearly changed (and much of the remaining disagreement between us has to do with the different use of the words in biology and cybernetics), such genes were specifically postulated, in a way which I believe was not just metaphorical, in the first full statement of the hypothesis. For instance, in discussing the possibility that organisms with frank detrimental effects on Gaia may gain a significant foothold on the biosphere, Lovelock (1979) writes: "In real life there must be taboos written into the genetic coding, the universal language shared by every living cell. There must also be an intricate security system to ensure that exotic outlaw species do not evolve into rampantly criminal syndicates."

3.3 *Other levels of selections: groups.* The program of successive reduction of the analysis of selection to lower and lower levels I outlined above is not the only way of looking at things, although it is now the most popular. There are ways of understanding the origin and maintenance of traits that benefit collections of related individuals as collections but are of no, or even negative, value to individuals. One of these, group selection, is illustrated schematically in Figure 4. Imagine a gene that causes its bearer to behave *altruistically*, to sacrifice its own reproductive interest (risk its own survival or curtail its own procreation) to the benefit of others. Warning calls in birds are a favourite example, but there are many others, including sex. Altruists will do poorly and ultimately become extinct in groups (interbreeding populations) in which there are also selfish individuals that lack the trait. But if

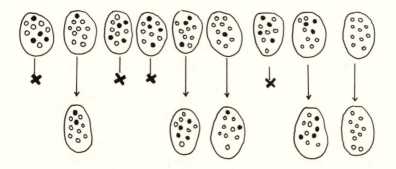

Figure 4 Group selection. A black dot indicates selfish individuals that are at a selective advantage within groups. A large X indicates group extinction, which is more likely in groups with more selfish individuals.

groups with altruists become extinct less frequently *as groups*, then the total number of genes for altruism, and the fraction of individuals that are altruists, can actually rise in the species as a whole, that is, in all surviving groups.

Thus group selection can *maintain* altruistic traits; it is a little more difficult to see how altruism could *arise* by this mechanism. Everything depends on the relation between rates of group formation and extinction and selective disadvantages to individuals. Population genetic modelling suggests that the conditions under which group selection can be important are quite restricted, and the current trend is thus to find underlying selfish reasons for apparently altruistic behaviour (Dawkins 1976; Maynard Smith 1976; Sober 1984). I doubt very much that one could buck this trend strongly enough to establish convincingly that altruistic traits, whose benefits are not specifically conferred on members of the same population but on the biosphere as a whole (and furthermore only after lengthy multi-multi-generational lag times), ever could arise by anything like group selection.

3.4 *Species selection: another model.* Gould, Eldredge, Stanley, and others who believe that the fossil record tells a story of *punctuated equilibria* (long periods of stasis in which few new species arise, and short periods of change in which speciation events are frequent) often also take into account selection at the still higher level of the species, and here I can get some encouragement (Stanley 1975).

Species selection takes the same conceptual form as selection at the level of individual organisms, with the process of speciation (separation into isolated populations) substituting for individual procreation, the process of species extinction for individual death, and the species itself for the individual, the entity that shows heritable variation in fitness. I go to Sober (1984) for an example:

We might visualize how this hypothesis works by a simple hypothetical example. Let's begin with two species, one winged and the other nonwinged, which initially have the same census sizes. Suppose that winged organisms survive and reproduce exactly as well as their wingless counterparts. But there is a difference: small colonies of wingless individuals become separated from the main population rather frequently; winged organisms, on the other hand, in virtue of their greatly mobility, very rarely form isolated subpopulations. If this system is allowed to evolve, we may later find a large number of rather small wingless species and a small number of rather large winged species. Wingless species speciate, whereas winged species grow. Notice that although wingless *species* have had more daughter species, wingless *organisms* are not more reproductively successful than winged ones.

This example is fascinating. The fossil record would show an increase in wingless species over time, and we might conclude from this evolutionary trend that winglessness was selected for slowly in geological time, that it was a better thing to be a wingless individual. But we would be wrong in this, and Sober argues that "species selection is *not fundamentally reducible* to organism or gene-level selection." I do not think we could deal with Gaia through species selection, but this at least allows us to think about evolutionary processes unique to higher levels of organization.

3.5 *Higher levels?* I am by now way out of my depth, but I should like to consider what may be still another level. My idea emerges from thinking about one of the most surprising recent results of molecular biology. Some genes of some organisms, we now know, are broken up into pieces scattered widely along chromosomes, each piece coding for only part of a protein. There is a tendency for this seemingly wasteful and cumbersome type of gene organization to be more common (to affect more genes more severely) in more complex "higher" organisms. Walter Gilbert (1978) suggested that the "function" of such organiza-tion was to promote the evolution of such complexity by allowing the formation of new genes from pieces of old genes – that this is what the organization was *for*. This is almost the same kind of *for* as Lovelock's, except that the long-term (very long-term) advantage is evolutionary adaptability, not global survival. The arguments against it are the same; evolution does not look ahead (Doolittle 1978; Darnell 1978).

And yet evolution seems to have looked ahead, because there is this correlation between the split-gene organization potentially conducive to the development of organismal complexity and complexity itself, both appearing to have increased over geological time. I have suggested elsewhere that split genes were present "from the beginning," that they are left over from the days of pre-cellular, self-replicating nucleic acids. Splits within genes are not, I believe, advantageous to individuals, and one would expect selection at the level of individuals to eliminate them gradually, in the interest of genetic efficiency (Doolittle 1978). However, since the story of life on earth is largely one of the increasingly complex and sophisticated exploitation of ever more diverse and complex habitats, split-gene organization will be maintained and may even be on the rise, because it makes the evolution of new, more complex, adaptations just a bit easier.

At this point, it is possible for a reductionist like me to get very confused. It seems almost as if I am suggesting that split-genes have been *selected* for, in response to a selection pressure coming from a geological trend in the direction of increased complexity, and that the level at which the selection has occurred is higher even than that of the species. My training tells me this is nonsense, and I am sure that a rigorously executed reductionist analysis could restructure this argument to remove all implications of evolutionary pre-adaptation, foresight, or progress. I am not sure, however, that some other way, some myth, which managed to incorporate the results of reductionist analysis into a larger and "existentially less irritating" model of the world, could not be constructed.

3.6 *Back to Gaia.* Although I am not willing to admit that Gaia could have arisen by anything like natural selection as we now understand it, the above discussion is an indication of willingness to admit that she might have arisen by some kind of selection that we do not yet understand. If it were real, this "selection" would have a paradoxical character, like the anthropic principle. That principle seems to me in an odd way quite similar in structure to the principle of evolution by natural selection. Organisms do things (have advantageous traits) because if they had not done them, they would not be here. The universe exhibits unusual properties (has life), because if it had not done, it wouldn't be observed – the effect seems to determine the cause. Again, I am sure I have not got this right, but I wonder if this is not because there is some larger and more profound principle here which my mind is just too poorly evolved to grasp.

For this reason it becomes very important to know if Gaia exists. Of course the biosphere affects the earth, and of course those effects feed back on life. But how complex are these feedback loops, and how stable? In other words, to what extent does the collection of homeostatic mechanisms appear accidental, like William Paley's stone, and to what extent does it appear to be the product of design, like his watch? And is that design so compelling that we could not deny the existence of a Designer? I hope so, and today I have the belief that it will be that compelling. But that faith is only articulated by what I have argued; it is certainly not derived from the data.

So we need some criteria for reality, some ways to distinguish accident from design.

4   Science and Faith

I suspect that most of us engage these questions for personal as well as professional reasons. I have used a reductionist methodology for a long time, and I am not sure any more that I have a sense of wonder about biology. I do not know that the methodology and the loss of wonder are actually related; my life has other strands. I do know that I should like a different model of the world, one which contained elements of mystery and counter-intuitive realities. I could adopt such a model from physics. Yet I am willing to explore the possibility that it could emerge from biology itself without denying the validity, at its own level, of reductionist materialism. I suspect that ultimately such a model would open out into something like a paradox, something which I could never properly explain to anyone. But it would be nice once again to get some delight from the law of the Lord.

WORKS CITED

Comfort, A. 1984. *Reality and Empathy: Physics, Mind and Science in the 21st Century*. Albany: State University of New York Press.

Darwin, C. 1859. *On the Origin of Species*. London: J. Murray.

Darnell, J.E., Jr. 1978. "Implications of RNA.RNA Splicing in Evolution of Eukaryotic Cells." *Science* 202:1257–60.

Davies, P. 1984. *God and the New Physics*. London: Penguin.

Dawkins, R. 1976. *The Selfish Gene*. Oxford: Oxford University Press.

Ditfurth, H.V. 1982. *The Origins of Life*. San Francisco: Harper and Row.

Doolittle, W.F. 1978. "Genes in Pieces – Were They Ever Together?" *Nature* 272:581–2.

Doolittle, W.F. and C. Sapienza. 1980. "Selfish Genes, the Phenotype Paradigm and Genome Evolution." *Nature* 284:601–3.

Gilbert, W. 1978. "Why Genes in Pieces?" *Nature* 271:501.

Lewontin, R.C. 1974. *The Genetic Basis of Evolutionary Change*. New York: Columbia University Press.

Lovelock, J.E. 1979. *Gaia: A New Look at Life on Earth*. Oxford, New York: Oxford University Press.

Maynard Smith, J. 1976. "Group Selection." Quart. Rev. Biol. 51:277–83.

Mayr, E. 1982. *The Growth of Biological Thought: Diversity, Evolution and Inheritance*. Cambridge: Harvard University Press.

Mills, S. and J. Beatty. 1979. "The Propensity Interpretation of Fitness." *Philosophy of Science* 46:263–86.

Monod, J. 1971 *Chance and Necessity*. New York: Knopf.

Orgel, L.E. and F.H.C. Crick. 1980. "Selfish DNA: The Ultimate Parasite." *Nature* 284:604–7.

Schrödinger, E. 1944. *What Is Life?* Cambridge: Cambridge University Press.

Shapiro, J. 1983. *Mobile Genetic Elements*. New York: Academic Press.

Sober, E. 1984. *The Nature of Selection: Evolutionary Theory in Philosophical Focus*. Cambridge: MIT Press.

Stanley, S.M. 1975. *Macromolecular Evolution: Pattern and Process*. San Francisco: W.H. Freeman.

Tillich, P. 1952. *The Courage to Be*. New Haven: Yale University Press.

Watson, A., J.E. Lovelock and L. Margulis. 1978. "Methanogenesis, Fires and the Regulation of Atmospheric Oxygen." *BioSystems* 10:293–8.

Williams, M. 1973. "The Logical Status of Natural Selection and Other Evolutionary Controversies." In *The Methodological Unity of Science*, ed. E. Bunge. Dordrecht: Reidel. 84–102.

P.J.E. PEEBLES

The Large-Scale Structure of the Universe

Introduction

The papers on astronomy and geology in this volume serve to illustrate the methods and results of the physical sciences, and to describe the nature of the physical environment within which life may form and other curiosities occur. Astronomy is not a very rich example of the first purpose because its observational basis is necessarily so limited – we cannot poke a star, or look at the universe from another place – and its inferences are therefore indirect and often controversial. Laboratory physics, however, provides a more convincing example of the way physics plumbs some aspect of reality. This problem with observations also makes cosmology a slim reed for the second, but as the connection is heavily advertised it is well to take a look at the situation.

The Main Results of Cosmology

We think we can work out a theory of the structure and evolution of the universe because of a very special situation.[1] It will be recalled that we are in an "island universe" or galaxy of some 10^{11} stars in a disc with diameter $\sim 10^5$ light years. There are about 10^{10} other galaxies within the reach of ground-based telescopes. These galaxies are found in groups and clusters, the largest of which are a thousand times the mass of our galaxy and a thousand times the diameter. When we smooth the distribution of galaxies over scales $\sim 10^8$ light years or larger we find that the clumpy character seen on smaller scales washes out, replaced with an accurately smooth isotropic mean distribution. The same effect is seen in counts of extragalactic radio sources and in the flux of X-rays emitted by clusters of galaxies, quasars, and the like.

This isotropy is something new. On every scale down to the smallest that has been explored we see rich phenomenology – quarks within nucleons within atomic nuclei and so on up to galaxies within clusters of galaxies. It certainly is conceivable that the hierarchy of structures sets in again on some scale larger than our horizon, and indeed that is what would be expected in the inflationary scenario Unruh describes: But we are presented with a curiously bland situation on the largest scales we can see, a universe evenly spread around us.

The galaxy distribution could be inhomogeneous but spherically symmetric about one point, so that isotropy would be observed from only a privileged few galaxies. In this case one would have thought that the privileged galaxies would look special. As far as we can see distant galaxies are much like ours, and as good as ours as homes for observers, so this is an unlikely situation. Thus the standard interpretation of the observed isotropy is that the universe is very close to homogeneous in the average over scales $\sim 10^8$ to 10^{10} light years.

Well before all this extragalactic phenomenology was known Einstein argued that a reasonable universe would be homogeneous and isotropic. He started from Mach's principle, which says that the motions we call inertial ought to be no more absolute than is any particular velocity. When we sense an acceleration it ought to signify an acceleration relative to the rest of the matter in the universe. General relativity theory admits solutions in which matter is in a bounded region outside of which space approaches the flat Minkowski space-time of special relativity. In such a solution a particle that moved away from the boundary of the mass distribution would retain its standard inertial properties, yet could be arbitrarily far from the matter that is supposed to define its inertia. De Sitter revealed that Einstein toyed with the idea of remedying this by assuming the occupied part of space is bounded by "hypothetical bodies" that somehow cause the exterior space to curl up into a singularity.[2] De Sitter objected that there is no observational evidence of these hypothetical bodies. Einstein then hit on a much more elegant solution: assume the distribution of mass is homogeneous, with no boundary.[3] I wish I knew what de Sitter thought of this; he surely told Einstein that the known universe of stars is bounded, consisting of our galaxy. However, Einstein persisted, and his insight or guess proved to be remarkably successful, setting a very bad example for later generations of cosmologists.

The background radiation observed from radio to sub-millimetre wavelengths provides remarkably strong evidence in support of this expanding universe picture.[4] The argument starts with the observation that if the universe really were expanding and evolving, then in the distant past the universe would have been a good deal more dense than it is now.[5] In a dense state matter and radiation rapidly relax to statistical equilibrium, producing blackbody radiation. It will be recalled that the spectrum of blackbody radiation is fixed by one parameter, temperature. As the universe expands this radiation is not lost – since the universe is supposed to be homogeneous there is nowhere for it to go – but it is cooled by adiabatic expansion, the temperature varying in proportion to $1 + z$, where z is the redshift factor defined above. Consistent with this picture, we are bathed in radiation isotropic to one part in 10,000 and with a spectrum that is very close to blackbody at a temperature of 2.75 K.[6] Most cosmologists take this to be almost tangible evidence, to be compared to dinosaur footprints, that the universe really did expand from a state a good deal denser than it is now, because no one has found a reasonable way to produce radiation with this distinctive spectrum in the universe as it is now.

Now another test is possible. Knowing the present temperature of the universe, and having at least a rough estimate of the present mass density, we can attempt to trace the thermal history of the universe back in time. If we extrapolate back to redshift factor $z \sim 10^{10}$ (to be compared to the largest observed redshift $z = 3.8$!) we get a temperature $T \sim 3 \times 10^{10}$ K, which is hot enough to decompose thermally atomic nuclei and to bring the abundance ratio of neutrons to protons to the statistical equilibrium value. As the universe expands and cools to $T \sim 3 \times 10^9$ K these neutrons and protons can radiatively combine to form deuterium (the heavy stable isotope of hydrogen), and deuterium can burn to form helium and heavier atomic nuclei. Laboratory values for the rates of all these reactions are well known, and general relativity fixes the rate of expansion and rate of cooling of the universe, so you can see that it is a straightforward if lengthy problem to compute the production of deuterium and heavier nuclei in the early universe.[7] In the conventional cosmology this nucleosynthesis depends on just one parameter, the present value of the mean mass density, for that fixes the density of matter at $z \sim 3 \times 10^9$, when the temperature dropped to the point that complex nuclei could form, and the matter density fixes the reaction rates. If the present density were $n_0 \sim 0.2$ protons per cubic meter then the computed abundances of deuterium, helium and the lithium isotope ^7Li all would agree with the estimates of the primeval element abundances derived from the oldest known stars. The required present density n_0 is not incompatible with direct estimates from galaxy masses and counts, as will be discussed below.

These calculations show that cosmology is starting to look like a mature science. The computation depends on a spectacular extrapolation from the present state of the universe, and for that reason most of us are hoping that the consistency of observed and computed abundances really is significant. The case will be stronger when we can measure the abundances of the light nuclei in a larger sample of galaxies, as it is expected will be possible with Space Telescope. Also, we are trying to reduce the uncertainty in the direct estimate of n_0 so that n_0 can be removed as a free parameter.

The value of n_0, as we shall soon see, figures in another very lively topic.

Instability, Initial Conditions, and the Mass Density of the Universe

The theories described above for the blackbody radiation background and nucleosynthesis at high redshift depend on the simplifying assumption of homogeneity. We have seen that this is a sensible approximation to the galaxy distribution averaged over large scales, but of course it is only an approximation: mass appears in lumps like stars and galaxies. What is the history of these departures from homogeneity? If general relativity theory is valid the universe is gravitationally unstable. That means the density fluctuations we see now will be larger in the future; that is, the hierarchy of structures will extend to larger mass scales. It also implies that the mass distribution must have been smoother in the past than it is now.

We have to pause to consider an apparent way out of these conclusions. The equations of motion of general relativity are invariant under reversal of the direction of time, so if we have a solution in which a smooth mass distribution is growing clumpy then by reversing the direction of time we get another solution in which a clumpy distribution is growing smooth. But this new solution is of no use, for it requires highly special initial conditions (the old final state), and the slightest error in the new initial conditions will grow, eventually dominate, and turn the decaying density irregularities back into growing ones.

The instability of the universe thus implies that the further back in time we trace the expansion the smaller the deviations from homogeneity. That is why the theories of the blackbody radiation background and of nucleosynthesis may make sense. Both involve long extrapolations back to epochs when the theory says the universe had to have been very nearly homogeneous and so a lot simpler than it is now.

Let us turn to another aspect of the instability of our expanding relativistic world model. The instability drives the initial balance between kinetic energy and gravitational potential energy toward dominance by one or the other of these two terms. This instability applies not only to local patches of matter but also to the general expansion observed within our horizon. It corresponds to the bifurcation between the cosmological models that approach free expansion with negligible gravitational deceleration and the models that eventually stop expanding and collapse back to an indefinitely dense state. Since our universe reached the present state with a rough balance of kinetic and potential energies in the face of this instability, kinetic and potential energies must have been finely balanced in the early universe. To state the case as dramatically as possible, let us suppose classical cosmology applies all the way back to the Planck time,

$$t_p \sim (hG/c^5)^{1/2} \sim 10^{-43} \text{ sec.}$$

This is the smallest time for which there is any hope that general relativity theory is a useful approximation. The present balance of potential and kinetic energies of expansion would imply that at the Planck time the two agreed to better than one part in 10^{60}. This is often quoted as an example of the spectacular fine tuning of the physical universe.

The point of all this discussion is that to account for the present state of the universe we must postulate initial conditions at high redshift that represent an exceedingly special and fine point-by-point balance. In thinking about what this might mean I find it helpful to review the role of initial conditions in other parts of physics. Turbulence is strongly unstable and dissipative, so the flow quickly forgets its initial state. That means we can get by with schematic initial conditions, and we can aim for a universal theory of turbulence. In the solar system dissipation has been important in some effects, like the alignment of the Galilean satellites in the equatorial plane of Jupiter. However, the motions of the planets are stable and very nearly without dissipation, so the present state largely reflects the details of the initial conditions; there is no universal theory for solar systems of the sort we have for turbulence. It is thought that the initial conditions for celestial mechanics were set by the evolution of the diffuse proto-solar nebula, whose properties in turn were set by events in the interstellar medium; and so on. It is to these processes that preceded celestial mechanics that we must look for the origin of such examples of fine tuning in the solar system as the peculiar alignment of the orbits and spins of the planets.

The evolution of density fluctuations in the expanding universe is thought to have had a very different character on small and large scales. On the scale of galaxies and smaller evolution has been strongly dissipative, so galaxies, like turbulence, are at best a murky reflection of what things were like in the beginning. The components of density fluctuations on scales greater than about 30 million light years are only weakly dissipative and weakly non-linear. Here present conditions recall initial conditions, as in the celestial mechanics of the solar system, and to end up with what we observe we must assume the expanding universe was assigned the special initial conditions of point-by-point balance.

What assigned the initial conditions? It is daring indeed to argue by analogy from a conjecture, but we are desperate, so let us proceed as follows. The solar system is a highly regular system that is presumed to be the result of the physics of the proto-solar nebula (though as Ovenden indicates, we will be hard pressed to demonstrate it). By analogy we might guess that the universe evolved from a state for which classical physics cannot provide a useful approximation. That idea is strongly supported by the fact that classical theory says the expansion traces back to a singularity. In a non-classical state analogous to the proto-solar nebula, physical processes might have tended to produce the observed initial conditions for classical cosmology. One possibility is the inflation scenario described by Unruh. Of course, inflation, or whatever else was at work, needs its own initial conditions, which we presume were set by something still deeper, like the bubbling of quantum space-time foam. One dream is that this hierarchy of physical processes is so entangled that we will come to see that statistically only one universe is possible, as in fully developed turbulence. Another dream is that the physical processes throw out lots of expanding universes, like so many planetary systems, some of which prove to be hospitable to observers like us.

There is a test that seems to have some bearing on these considerations. Suppose a deeper physical process did produce the present fine balance out of chaos. Since the balance is good enough to assure homogeneity of the galaxy distribution on scales $\sim 10^8$ light years it must have been good enough to assure balance on the scale $\sim 10^{10}$ light years of our horizon. But on the latter scale we are talking of the general expansion of the observed universe, and the balance implies that from the observed expansion rate we can predict the gravitational energy and hence the mean mass density in our horizon. It amounts to a few protons per cubic meter. So we have a test, but unfortunately estimates of what the density really is are exceedingly difficult, and I can only report the direction of current observations and ideas.

Direct estimates of masses and abundances of galaxies indicate that if most of the mass of the universe is in and around galaxies then the potential energy is about 10 per cent of the kinetic energy of expansion.[8] About the same result follows from the computation I mentioned earlier of the production of light elements in the early universe (see n 7). If these indications prove to be correct it will be a very interesting problem, for it is hard to imagine a physical process that would strike the wanted balance of initial conditions locally (on the comoving length scale $\sim 10^8$ ly) while missing by a factor of ten on the scale of our horizon (comoving length $\sim 10^{10}$ ly).

A way out that is attracting a lot of interest is to assume the global balance really is present and we have missed it because some 90 per cent of the mass is in an undetected "dark" form. We must postulate that this dark mass has no strong interaction, so it does not interfere with nucleosynthesis, and that it is spread between the galaxies, so we miss it when we add up galaxy masses. Particle theorists have supplied us with a host of possible dark matter particles.[9] There are, then, grounds for optimism and caution; the very variety of candidates shows that most of the suggestions for dark matter particles must be wrong for otherwise we would get too much dark matter. Another possibility is that the dark mass is manifest via Einstein's cosmological constant Λ. This requires a curious coincidence, that the value of Λ just happens to offset the factor of 10 deficit in the density of seen matter.[10]

All this is very unsatisfactory but certainly not grounds for despair. It merely shows the usual fumblings of an immature but lively science.

What Does It All Mean?

The expanding world model has led to some notable successes and curious puzzles. The puzzles need not signify problems with the overall picture. We have had to get used to the idea that a theory that is experimentally very successful, and so surely mirrors reality, can also exhibit pathologies that we hope are only accidents of the particular approximations of the theory. But still, the ratio of successes to puzzles is uncomfortably small in cosmology, a sign of the immaturity of the subject.

The curious regularities and fine tunings of cosmology have been used in two ways, as guides to a new physics[11] and as indications that conventional science may be missing part of the story. Both are clearly analysed by Leslie.[12] The first has a long history of successful applications in other fields and surely is worth a try here. The second seems to me somewhat less convincing. The modest successes of extragalactic astronomy help reinforce our belief that it is useful to assume nature operates by rules that are simple (according to sufficiently flexible criteria) and that we can hope to discover. Wigner has given the clearest scientific expression of how wonderful it is that nature should operate this way.[13] But given that it does we must observe systematics. And in any reasonably large space of possible systematics whatever is observed has an exceedingly low probability. Thus the existence of curious regularities *per se* is nothing new. I particularly admire Hacking's statement of this point. The specific regularities of cosmology are beneficial to our form of life, but I do not see how we can attach deep meaning to this inference until we have some idea of how many forms of life nature can come up with. In that connection, we should remember that the fact that nature seems to choose to operate by rules does not impoverish the phenomenology; we have seen time and again how nature can surprise us without (so far as we can tell) bending the rules. Thus I suggest that you be wary of cosmologists who come to you bearing curious puzzles.[14]

NOTES

1 For a more detailed survey of these results see Section 1 of my *The Large-Scale Structure of the Universe* (Princeton: Princeton University Press, 1980).

2 W. de Sitter, *Monthly Notices Roy. Astronom. Soc.* 77 (1916): 181. 3 A. Einstein, *S.B. Preuss. Akad. Wiss.* (1917): 142.

4 For a review of redshift observations see S. Djorgovski and H. Spinrad, *Astrophysical Journal*, in press, 1985.

5 See, for example, M.S. Roberts in the proceedings of IAU Symposium 44, *External Galaxies and Quasi-Stellar Objects*, ed. D.E. Evans (Dordrecht: Reidel, 1972), 33.

6 See, for example, G.F. Smoot et al., *Astrophysical Journal Letters* 291 (1985): L23, and references therein.

7 For the most recent computation and observational tests see J. Yang et al., *Astrophysical Journal* 281 (1984): 493, and references therein.

8 The measurements that finally convinced me of this are described by M. Davis and P.J.E. Peebles, *Astrophysical Journal* 267 (1983): 465, and by A.J. Bean et al., *Monthly Notices Roy. Astronom. Soc.* 205 (1983): 605.

9 For a review of the dark matter candidates see H. Pagels in the proceedings of the Eleventh Texas Symposium on Relativistic Astrophysics, *Ann. N.Y. Acad. Sci.* 422 (1984): 15.

10 P.J.E. Peebles, *Astrophysical Journal* 284 (1984): 439.

11 For examples of this approach see R.H. Dicke and P.J.E. Peebles in *General Relativity*, ed. S.W. Hawking and W. Israel (Cambridge: Cambridge University Press, 1979), 504; and A. Guth in the proceedings of the Eleventh Texas Conference on Relativistic Astrophysics, *Ann. N.Y. Acad. Sci.* 422 (1984): 1.

12 J. Leslie, "Anthropic Principles, World Ensemble, Design," *American Philosophical Quarterly* 19 (1982): 141.

13 E.P. Wigner, *Communications on Pure and Applied Mathematics* 13 (1960): 1.

14 Research for this paper was supported in part by the U.S. National Science Foundation.

MICHAEL W. OVENDEN

Of Stars, Planets, and Life

Prolegomena

In the middle of the sixteenth century, artists were also scientists. To a young lad growing up at that time, it was the artist who seemed most fully to express the promise of human genius, and his greatest ambition would have been to become a famous architect, painter, or sculptor – and man, forever making God in his own image, would seek design in nature as evidence for a Divine Artist. The recognition of "design" was therefore in the nature of an aesthetic judgment.

On the other hand, to a young lad growing up in Scotland in the middle of the nineteenth century, it was not the artists but the engineers who were the "astronauts" of his day, pushing forward the boundaries of human experience. While a machine may indeed have an aesthetic beauty, the evidence from design for a Divine Engineer would now have to rest upon judgments other than the aesthetic, which had come to be seen as subjective.

One such lad was William Thomson, later to become Baron Kelvin of Largs, one of the greatest physicist-engineers of all time. Kelvin (amongst others) sought a theoretical basis for the empirical science of thermodynamics, which had grown in response to the needs of the developing steam-powered railways. The answer came in statistical mechanics, in which the behaviour of a gas was understood in terms of the motions and collisions of innumerable elastic atoms like tiny billiard-balls, heat being but the random motions of such atoms. The connection between statistical mechanics and thermodynamics led to the definition of an "ordered" state of matter as an "improbable" state. For example, if a ball is thrown randomly into a box, there is a 50 per cent chance that it will fall into the left-hand half of the box. If five balls are so thrown, the chance that all would fall into the left-hand side would be $(1/2)^*(1/2)^*(1/2)^*(1/2)^*(1/2) = 1/32$, already a moderately small number. Clearly the chance that all of the atoms in a match box (many billion of them) would be, "by accident," in one half of the box is indeed astronomically small. The state with all the atoms in one half of the box is said to be a highly ordered state, the order being measured by the improbability of that state arising by chance.

Thus in the nineteenth century, the aesthetic notion of "design" was replaced by the quantitative definition of "order" – God had become the Divine Orderer. It is in this sense that P.C.W. Davies, in his book *The Accidental Universe*, is able to describe the "improbability" of the cosmos as "its extraordinarily contrived appearance." I shall stay with this reduction of "design" to "order" for most of my paper, but I shall return to some fundamental problems with such a reduction at the end.

The history of the matter in the universe is its successive coagulation, under the attraction of self-gravity, through a hierarchy of condensations (which I shall discuss in a moment) into objects of stellar and planetary size. The first stage in this hierarchical sequence is represented by the galaxies, objects whose masses are of the order of some hundreds of thousands of millions of times the mass of a typical star like the Sun. There are some problems in understanding the formation of galaxy-sized condensations within an expanding Universe, but Professor Unruh discusses these, and I shall take condensations of the mass of a typical galaxy as given. How does a galaxy fragment further into stars, and what happens to these stars?

The Evolution of a Star

A star like the Sun begins as a vast condensation in interstellar matter. As it contracts under the mutual gravitation of each part upon every other part, it assumes a spherical shape, and as the matter of which the condensation is made falls in towards the centre of the sphere, the speeds of infall of the particles increase, and collisions between the particles gradually randomize the velocities. In other words, the contracting sphere gets hotter as it gets smaller. To understand the evolution of such a condensation, we have to use computer calculations to follow the consequences of the known laws of physics (which explicitly we assume to apply). We find that the centre of the sphere gets hotter more quickly than the outer regions.

At a rather well-defined moment in the history of the star (which, for a star like the Sun, occurs some few million years after the condensation first forms), the centre of the star is so hot (about 15 million degrees

Celsius) that the energy of motion of the atoms is great enough to overcome the forces that maintain the integrity of atomic nuclei. Hydrogen nuclei fuse together to form nuclei of helium, and in this process energy is given out. The emission of energy halts the contraction, and the star remains for a significant part of its lifetime in quasi-equilibrium, generating energy in its core by such thermonuclear reactions, and getting rid of the energy at its surface in the form of electromagnetic radiation (X-rays, light, radio waves, etc.). During this phase of its career, the star is said to be on the "main sequence."

However, a star cannot stay for ever on the main sequence, because it is steadily converting its available hydrogen into inert helium. The Sun has been on the main sequence for about half of its main-sequence lifetime of some 10,000,000,000 years.

As the star uses up its hydrogen in its central regions, the core of inert helium (while growing because of the addition of new helium), also contracts like a "star within a star" – and as it contracts, it heats up. Meanwhile the star itself becomes more luminous; but, as it also grows much larger, the surface of the star becomes cooler. The star has become a "red giant." The star Betelgeuse in the constellation of Orion is an example. Although its mass is only about twenty times the mass of the Sun, it is so tenuous that if placed where the Sun is, Betelgeuse would engulf the orbits of the Earth and Mars.

The evolution of stars whose masses are different from that of the Sun is qualitatively the same, but the rates of evolution are different. The more massive the star, the faster its evolution, and the shorter its lifetime on the main sequence. The visible stars vary in mass from about 1/100 to 100 in relation to the mass of the Sun.

Our confidence in the basic reliability of our theory of stellar evolution comes from looking at the properties of clusters of stars. In a group of stars formed together, at the same time, the most massive stars will have contracted to the main sequence, spent their main-sequence lifetimes, and evolved into the giant phases before the least massive stars have yet reached the main sequence. Statistics of stars of different luminosities and temperatures in actual star-clusters, enable us to calculate consistent ages for the clusters. In particular, clusters whose brightest stars are blue must be young clusters (the Pleiades, for example, are about 50 million years old), while clusters whose brightest stars are red must be old (the globular clusters are about 10 thousand million years old).

The subsequent evolution of a star is complicated. It involves the generation of elements heavier than helium by thermonuclear reactions between the helium nuclei. It may also involve instabilities; if massive enough the star may even explode as a supernova. The final end of a star is as an exhausted "cinder," which I shall discuss briefly later.

Can our present-day physics enable us to understand how interstellar matter can condense into units whose masses are star-like? Imagine a condensation beginning to form in a diffuse cloud of matter, which (following our knowledge of the chemical composition of the universe) must be mostly hydrogen. According to the law of gravity, for a hydrogen atom at the edge of the condensation, the surrounding material has no net effect, while the condensation acts as though its mass were concentrated at its centre. Left to itself, the atom would tend to fall into the condensation. However, because the gas out of which the condensation is forming has some thermal energy, the atom will usually have some speed relative to the centre of the condensation. If this speed is less than the escape speed from the condensation, the atom will start to fall in an orbit that will take it into the condensation. If, however, the atom has a speed greater than the escape speed from the condensation, the atom (while still being attracted to the condensation), will move in a curved orbit that takes it away from the condensation (just as a rocket sent up from the Earth with less than the escape speed from the Earth will fall back to the ground, while a rocket ejected with a speed greater than the escape speed will leave the Earth). In any "hot" gas, there will be a range of thermal speeds; the higher the average speed, the higher the temperature. Thus the mean thermal speed must be less than some fraction of the escape speed while at the same time being a function of the temperature. The higher the temperature, the greater the minimum mass for a stable condensation, while the greater the density of the material in which the condensation is occurring, the less the minimum mass. (The exact equations are given in Appendix 1; the formula for the minimum mass of a stable condensation is known as Jeans' Criterion.)

In Table 1, Jeans' Criterion is applied to a number of situations, with physical parameters such as temperature and density being assigned from observation. I consider first condensation from material whose density corresponds to that of matter between the galaxies, as we find it today. The temperature of such intergalactic matter is simply taken as that of the cosmic background radiation. Jeans' Criterion shows that a mass of about a million suns would be stable against temperature dissipation. This is about the mass of a typical globular cluster at the edges of our galaxy. We must look with caution upon this agreement, since the globular clusters are almost as old as the Universe itself, and were therefore formed when the background temperature was much

Table 1 Jeans' Criterion

	μ	T	ρ	M
Intergalactic	1	3	$1 \times 10^{+7}$	$1 \times 10^{+6}$
Galaxy	1	3.6	$5 \times 10^{+1}$	$6 \times 10^{+2}$
Interstellar cloud	1	$1 \times 10^{+2}$	10	$2 \times 10^{+4}$
Orion nebula	1	$7 \times 10^{+3}$	$6 \times 10^{+2}$	$1 \times 10^{+6}$
	1	$1 \times 10^{+2}$	$6 \times 10^{+2}$	$2 \times 10^{+3}$
Globule (99.8% dust)	500	$1 \times 10^{+3}$	$6 \times 10^{+3}$	2
	500	$1 \times 10^{+2}$	$6 \times 10^{+3}$	0.1

μ = mean molecular weight, in units of mass of hydrogen atom
T = kinetic temperature in degress Kelvin
ρ = density, in hydrogen atoms/cc (equivalent)
M = mass of condensation, in units of the mass of the Sun

higher. While this factor tends to be offset by the much higher density of matter in the past, it was (according to simple theory) much more difficult to form condensations in the remote past. However, we do find a few young globular clusters between galaxies, and so our theory seems to be satisfactory at this stage.

Now we turn to the matter between the stars in our, or some similar, galaxy. Because of the heating effect of general starlight, its temperature will be somewhat greater than that in intergalactic space. Jeans' Criterion indicates that condensations whose mass is greater than about 600 suns would be stable. This prediction is in good agreement with observation, since we find interstellar matter condensed into clouds whose mass is of this order; for example, the Great Nebula in Orion, a location of many very young stars, has a mass of several hundred stars.

Can such an interstellar cloud fragment into smaller condensations of stellar mass? It seems that this must happen, since we find very young stars in such clouds, and furthermore the distribution of young stars (even when the parent clouds have been used up) is patchy, suggesting that stars are formed in associations of several hundred objects, which associations are themselves not stable and which dissipate.

Here, however, we run into trouble. The density and temperature of a typical interstellar cloud are such that only a condensation whose mass is a few tens of thousand of stars could condense – and this, of course, requires much more matter than is actually in the cloud. In a denser cloud, such as the Orion Nebula, the radiation from young stars already formed so heats the gas that the minimum mass for further condensation is up to about a million suns.

Yet stars do form! The solution seems to lie in the existence of dark absorbing interstellar matter, which is easily visible against the bright glow of the background nebula, especially in such interstellar clouds as the Orion Nebula. This absorbing matter is in the form of grains (primarily of carbon). Each grain has a mass of the order of many million hydrogen atoms. Because of their high masses, at a given temperature the speeds of the grains will be very much less than the speeds of the hydrogen atoms. Not only will the grains be more easily captured by a condensation, but also the presence of the grains will tend to cool the region by removing (by collision) the high-speed hydrogen atoms, and radiating away energy as infra-red rays. It seems quite possible to form condensations of stellar mass in an interstellar cloud which is made up primarily of grains.

In nebulae such as the Orion Nebula, then, we do see more or less spherical blobs of dark absorbing matter, called "globules," and conditions inside a globule seem suitable for forming stars. The only trouble is that our present estimates of the mass of a typical globule fall far short of the mass of a star like the Sun. Nevertheless, given enough material in granular form, the condensation of stars seems possible.

But how are the grains formed? Some of them certainly seem to come from evolved cool red giant stars, which are (as it were) smoking like a poorly-lit candle. This does not help us much in forming stars in the first place. There may be other factors that have been left out of account. In other galaxies, and probably in our own, stars seem to be formed not at random, but in patterns. Bright blue stars (the young ones) form characteristic spiral patterns, and one theory is that such spiral patterns represent an ordered arrangement of the orbits of stars in the galaxy, and such an ordered arrangement of orbits will produce a pressure on the gas in the spiral arms as stars and interstellar matter move into (and then out of) the arms. Such a pressure could assist the formation of stars; it has not been taken into account in the Jeans' Criterion.

Of course, we must then ask how such an ordered arrangement of orbits can occur: could we apply the "argument from design" to such an arrangement? Our present investigations suggest that such a pattern is formed by a galaxy which has been perturbed by the gravitational effect of another passing galaxy, rather as the surface of a drum vibrates in a definite pattern when it is hit by a drumstick. In the case of our galaxy, the culprit seems to be the Large Magellanic Cloud, a small satellite of

Table 2 Maximum Sizes for Cold Spheres

	M	R	ρ
Molecular	$2 \times 10^{+28}$	$8 \times 10^{+3}$	10
Atomic	$2 \times 10^{+30}$	$7 \times 10^{+4}$	1
Electron degeneracy	$3 \times 10^{+33}$	$9 \times 10^{+3}$	$9 \times 10^{+5}$
Nuclear	$4 \times 10^{+33}$	15	$3 \times 10^{+14}$
Stars and Planets			
Earth	$6 \times 10^{+24}$	$6 \times 10^{+3}$	5.5
Jupiter	$2 \times 10^{+30}$	$7 \times 10^{+4}$	1.4
Sun (hot)	$2 \times 10^{+33}$	$7 \times 10^{+5}$	1.4
Sirius B (White Dwarf)	$2 \times 10^{+33}$	$15 \times 10^{+3}$	$1 \times 10^{+5}$
Pulsar	$3 \times 10^{+33}$	15	$2 \times 10^{+14}$
Black Hole	$<5 \times 10^{+33}$	0 ?	∞ ?

M = mass of sphere in grams
R = radius of sphere in km
ρ = density in gm/cc

our galaxy which can be seen in the southern hemisphere as a fuzzy patch next to the Milky Way. Another possible source of pressure is radiation pressure from young hot stars already formed. But the question is controversial and far from clear.

In short, we find some difficulty in forming stars under conditions that we find in our galaxy today, but the difficulty does not seem insuperable; our theoretical ingenuity seems to be adequate to finding a reasonable explanation that does not place excessively stringent demands upon the properties of a galaxy. I turn now, briefly, to consider what happens to a star when it exhausts its nuclear fuel, and becomes a cold body.

I am sitting on a table. What prevents the table from collapsing and letting me fall upon the floor under the influence of the gravitational attraction of the Earth? Assuming that the table has been soundly made, and does not in fact collapse under duress, it is because the forces between the molecules of which the table is made are stronger than the force of my weight. This means that for any given force opposing the self-gravity of a spherical body, there is a maximum size for that body. Try to make the body larger, and the gravitational force will overcome the resisting force. For example, consider a sphere of ice, far away from the heating influence of a star. It is well known that ice will melt under pressure, when the pressure exceeds the molecular forces that maintain the crystal structure of the ice. The bigger the block of ice, the greater the internal pressure produced by self-gravity. There must be an upper limit to the size of a cold block of ice. There is: it is about 500 km radius (the equations are given in Appendix 2).

Now we can apply these ideas to bodies made of stronger material. The strongest molecular forces, more typical of rock than of ice, can sustain a sphere of about 8500 km radius, and a mass of about twice the mass of the Earth (see Table 2). Above this mass, the molecular bonds are broken, and the material becomes a collection of atoms. The atomic forces are associated with the shells of electrons surrounding the nuclei of atoms–they are the chemical forces, and can sustain a mass of about 200 earth masses. Jupiter is already somewhat more massive than this, indicating that there must be some additional resisting force, perhaps residual thermal energy – a speculation consistent with the observation that Jupiter emits more energy than it receives from the Sun.

After the electron shells have been crushed out of existence, the material of which the object is made now consists of nuclei and electrons moving about freely, unattached to each other. A new force now comes into play, one not known in classical physics. According to quantum theory, it is impossible to have two neighbouring electrons in the same energy state. As the mixture of electrons and nuclei is squashed more closely together, it becomes increasingly difficult to squash the material still further, because the electrons have less "space" into which they can go. This resistance to further squashing is called "the force of electron degeneracy." It sets a limit of about 1.4 times the mass of the Sun, and a radius of about 9000 km. I have been calculating the sizes of cold spheres. The Sun is much larger because it is hot. However, when it does exhaust its nuclear fuel, some 7 billion years in the future, it will shrink to a diameter of only a few thousand kilometres. Such objects, or "white dwarfs," have been observed.

Still, the force of electron degeneracy is not infinite. As we increase the mass of our cold sphere even further, the electrons have to find somewhere to go; they go into the atomic nuclei, combining with the atomic nuclei to form neutrons. The forces that hold the nuclei together can sustain a mass of perhaps three times the mass of the Sun, at a radius of a few kilometres. We observe such neutron stars as "pulsars," objects which emit regular bursts of radio radiation.

Even the nuclear forces are finite, however. According to ideas current in physics, there are no forces stronger than the nuclear forces. A star ten times the mass of the Sun must either get rid of almost all its mass (by exploding as a supernova?) or else contract indefinitely into a point-singularity of mass which we call a "black hole."

As we look at the masses and sizes of bodies in the universe, we see a reflection of the forces that sustain them. Terrestrial planets are sustained by molecular forces; the giant planets like Jupiter and Saturn are sustained primarily by chemical forces; white dwarf stars are sustained by degeneracy forces; neutron stars by nuclear forces.

The Sun is important, indeed vital, to our existence. Can we then apply the "argument from design" to the existence of sun-like stars? To translate this question into quantitative, rather than aesthetic, form: are stars improbable? Only an astronomer could ask such a stupid question! There are tens of thousands of millions of stars in our galaxy alone, and thousands of millions of galaxies. Clearly stars are not "improbable" in the universe as we find it. But does this mean that (in Davies' terminology) the universe has to be "extraordinarily contrived" for stars to be possible? At the end of my presentation, I shall deal with some of the logical problems associated with this view. At present, I shall assume that it means something like this: "In all the Universes that I could imagine, in which fraction of them would the formation of stars take place?"

To restrain the flights of my fancy, and to give an example of how one might try to answer such a question, I will restrict myself to considering the possible universes that would arise if the "constant of gravitation," G, could change its size relative to the atomic and nuclear forces. I am not (I hasten to add) supposing that in our actual universe, G varies with time. A similar hypothesis has been made, although it goes against current observations. I am simply imagining hypothetical universes, in each of which G is unvarying, but in which the ratio of G to the constants of nuclear and atomic forces is different from one hypothetical universe to another.

Figure 1 shows the results of such imaginative "speculations." The equations on which Figure 1 has been based are given in Appendix 3. To cover a sufficiently wide range of situations, Figure 1 is plotted with logarithmic axes, in which a shift of $+1$ represents a multiplication by 10, a shift of -2 a division by 100, and so on.

First, in order that we may have a stable condensation we may call a star, the central temperature must be high enough for thermonuclear reactions to take place. Since we have allowed that the constant of gravitation or G varies, but that the atomic and nuclear constants do not, the required temperature may be found from theoretical and observational investigation of the universe as we find it. Second, the star should not be too massive, or the fraction of its internal energy that is in the form of radiation will be great enough to "explode" the star

However, from the point of view of the existence of life, two other criteria must be considered. Life, even vaguely like life as we know it on the Earth, could only exist on a planet moving in a moderately stable orbit about a star whose luminosity is of a magnitude to maintain, on the planet, the right mean between the temperature at the surface of the star, and the cold of interstellar space. Somewhat arbitrarily, I have taken the luminosity limits to be from 1/10 to 10 times the solar luminosity.

In addition, I have assumed that the development of "intelligent" life requires a main-sequence lifetime for a parent star of the order of the present age of the Sun on the main sequence. In drawing the line marked t in Figure 1, I have (again arbitrarily) assumed the minimum main-sequence lifetime to be 1/3 that of the Sun.

If these criteria are accepted, the shaded area in Figure 1 shows the range of possible gravitational constant and stellar mass for which the development of intelligent life would be possible.

Two features of this "biogenic" region should be noted: (a) the range of stellar mass is greatest for that value of G corresponding to the observed value of the constant of gravitation; and (b) the shaded region is finite.

Property (a) is simply a result of having taken the standard of reference for the required lifetimes and luminosities to be the corresponding values for the Sun. The quantitative significance of property (b) is less clear. As was pointed out, the adopted range of luminosity for the central star of our hypothetical planetary system was somewhat arbitrary, because changes in luminosity can be compensated (to a large extent) by changes in the distance of the supposed biogenic planet from the star. Furthermore, if we allow changes in the relative magnitude of the atomic and the nuclear "constants," we could reduce the separation of the *Radn.* and *Temp.* lines; but we could not (according to our present-day understanding) reduce it to zero. It seems, then, that the biogenic region is finite. Is it also small? In other words, is the required relationship between gravitation and the atomic constants "improbable"? No objective answer to this question can be given, since the total area of Figure 1 is determined only by the limitations which have been arbitrarily imposed on the permissible range of G from 1/1,000,000 to 10,000 times its actual value. For example, is there any reason why G should not have been 1/1,000,000,000,000,000 of its actual value?

I shall return to this point later. Meanwhile, I shall proceed to discuss further the nature of the restrictions imposed on the universe such that biogenic planets may exist.

The Formation of Planets

The theory of the formation and evolution of stars is capable of critical testing because we observe a large number of stars of different masses and in different stages of evolution. Pending the discovery of planetary systems about other stars in sufficient numbers for statistical arguments to apply, the theory of the formation of planetary systems must remain to a considerable extent speculative.

Following a similar classification in geology, we may divide theories of the origin of planetary systems into "uniformitarian" theories which see the formation of planets as simply a stage in the hierarchical condensation process already described, and "catastrophic" theories which see a planetary system arising as a result of a special concatenation of happenstance, more or less improbable. On a uniformitarian theory, we would expect every star to have a planetary system during at least part of its lifetime; the probability of planetary systems on a catastrophic theory would depend upon the improbability of the particular congeries of circumstances demanded by the theory.

Copernicus, in his heliocentric theory, transferred the frame of reference of motion from the Earth to the background of the so-called "fixed stars." As the Copernican Revolution worked its way through European science, the conviction arose that the Sun was but one among myriads of similar luminous objects, the stars. If the Sun has planets, must not the other stars also have planets? Most popular astronomy books of the eighteenth century assumed automatically that every star had its family of inhabited planets.

At the end of the eighteenth century, Laplace described his famous "nebula theory" of planetary formation. It was a uniformitarian theory. As the condensation which was to form the Sun contracted, it increased its angular velocity of rotation by the conservation of angular momentum, similar to the way a skater who rotates with arms outstretched increases her rate of rotation as she brings her arms downwards. As material at the surface of the condensation reached circular velocity, it would become detached as a ring, which subsequently condensed into a planet. This would happen in successive stages, giving rise to several planets.

Laplace's theory was published only in a popular book on astronomy. In fact, analysis shows that the theory, in the form described, is not viable. In the first place, the material in such a ring would be hot, and would evaporate before it could form a planet. In the second place, while over 90 per cent of the angular momentum of the solar system resides in the orbital motions of the planets, the Sun itself is rotating so slowly that even if all the angular momentum of the planets were given to the Sun, it would still be rotating far too slowly to break up in the manner required by the theory.

To overcome the angular momentum problem, it was proposed in the early years of this century that the angular momentum of the planets derived from a passing star. In the best developed version of this theory, by Jeans and Jeffreys, the material from which the planets condensed was pulled from the Sun by the tidal effect of a star passing close by. There is still the problem of why the tidal action did not set the Sun spinnning more rapidly, an objection that was overcome by Lyttleton, who postulated that the Sun was a member of a binary system, and it was from the former companion of the Sun that the planetary material was derived, the tidal encounter also breaking up the binary pair.

Now it is not at all unreasonable to postulate an originally binary nature for the Sun, for 50 per cent of stars seem to be binary or multiple. Nevertheless, encounters between stars as they orbit in the Galaxy are very rare, and so the number of stars in the Galaxy that might be expected to have planets would have been very small, perhaps only a handful. The tidal theories must be considered to be "catastrophic."

Another problem remained for these theories. If the planetary material came from the surface of a star, it would have evaporated before it could condense into planets. After the Second World War, opinion tended to veer towards a new version of the uniformitarian "nebula hypothesis," in which electromagnetic forces supplemented gravitation during those phases of the contraction process for which the material was ionized, and in which it was recognized that a substantial fraction of the original circumsolar material was lost to the system, our planets being formed from the residue. With such a loss of material, there was also a loss of information, and we may never be able to work back from the present state of the solar system to its origin with any certainty.

Some version of the nebula hypothesis is generally favoured today. However, tidal theories have also come under re-examination, since it is recognized that stars form in groups in interstellar matter. During such formation, the separate condensations formed will tend to be rather evenly spaced, and with small velocities. They will therefore initially fall in towards one another, and suffer close encounters. Computer simulations show that the result of such encounters is to form binary stars, and what are called "runaway stars," which have an abnormally high space velocity; such runaway stars have been observed. Far from being unusual, it would seem that every star might be expected to suffer close encounters in the early stages of its evolution. Computer calculations of the result of such encounters indicate that planets can condense as a result of tidal action.

With only one known planetary system, it is very difficult to judge these different theories critically, and there are many divergent opinions on the subject. Can we not help solve the problem by looking for planetary systems around other stars? Unfortunately, currently available techniques are inadequate to the task. If we took our best telescopes to the nearest star, and looked at the Sun, we could not see even Jupiter directly, because it would be so much fainter than the Sun. Modern techniques of image analysis might be able to provide observations taken from orbiting observatories above the Earth's atmosphere, but the task would be time-consuming and expensive. Attempts have been made to detect invisible planets by their gravitational perturbation of the motion of their parent stars. Had I been writing this paper ten years ago, I would have said that for some dozen stars, such invisible planets had been detected. But the very small shifts in star-positions that indicate such perturbations have not been confirmed, except for one case, Barnard's Star, which is six light-years away, and one of our nearest neighbours.

However, observations in the infra-red from satellites have shown that a large number of stars have circumstellar dust in some form or another, and in at least one case (the star beta Pictoris) refined optical observations have demonstrated that the dust is in the form of a flattened disc, such as we would expect a nascent planetary system to be.

While evidence on the question of the frequency of planetary systems is still being accumulated, there is much more general agreement on the mode of planet formation. First there is the dust that seems necessary if stars are to form at all. If the dust grains are primarily carbon, with much hydrogen around, we might expect to find some hydrocarbons formed, for hydrocarbons are notoriously sticky. Thus we can picture the grains sticking together to build solid planets by "cold accretion," as it is called. As the condensations get larger, the energy of infall of one condensation on another under their mutual gravity gets larger. The energy of collision may be sufficient to melt the surface, and if the protoplanet is big enough, gases may be released to form an atmosphere. Over the last decade, observations from space-craft have shown that cratering is a common feature of planets and satellites, and this cratering is seen as evidence for the cold accretion hypothesis.

In short, given a disc of dust, the formation of planets seems necessarily to follow. What is the chance that planets will form that are suitable for the development of "intelligent" life?

Life on Other Worlds

As we have seen, the nature of the atomic and molecular forces limits the size of cold bodies to certain ranges, one of which corresponds to terrestrial-type planets. How "improbable" is it that the molecular and atomic forces would have just the right strength? We can play the G-varying game again. In this case, we find that as we vary G, the acceleration at the surface of a typical terrestrial-type planet (i.e., one constrained by molecular forces) will vary as the square-root of G. But I know no law of biology or physics that says that living organisms are possible only within certain ranges of gravitational acceleration. I see no objection to conceiving of life in a large gravitational acceleration that is almost two-dimensional, with a "cerebral cortex" in the form of a flat sheet, rather than in packed convolutions. Of course this is science fiction, but once more it points to the fact that the probability that we assign to a singular event depends upon our current theoretical understanding.

Often the argument in favour of the existence of intelligent life in the universe runs something like this: "There are hundreds of thousands of millions of stars in our galaxy alone, and millions of other galaxies. It is inconceivable that life developed only on the Earth." But why is it inconceivable? Nature is prodigal. For every seed that matures on Earth, there are many thousands of millions of pollen grains that do not lead to development. If we supposed, for the moment, that the Universe was indeed created just so that we might exist, this offends no principle of logic of which I am aware; it merely offends the sensibilities of a civilization that has become belatedly conscious of its own wastefulness.

It is my experience that biologists are more sceptical of the widespread occurrence of life than are astronomers. In fact, I think that there are at least two reasons for suspecting that, whether potentially life-bearing planets are common or not, life may be rare.

In the first place, it is a singular fact that all life on Earth has the same stereochemistry. A complicated molecule may often exist in two distinct forms, or isomers, one of which is the mirror image of the other. Now as far as we know, chemical reactions occur between molecules at the same rate and with the same facility, whichever form the molecules take. Yet the three-dimensional form of molecular structures plays a vital role in the activity we call life. Alice, in *Through the Looking Glass*, speculates on whether her cat could live on looking-glass milk. The cat could not. So, life-systems of opposite stereochemistry would not compete for food. The fact that all life on Earth has the same stereochemistry suggests that it developed from a single original examplar.

In the second place, from the cosmic point of view, the planet Mars is almost identical to the Earth. It is a little smaller, but massive enough to retain an atmosphere. This atmosphere, although very thin by terrestrial standards, does contain some oxygen and some water vapour. The temperature on the surface of Mars rises above the freezing point of water at the Martian equator. While large bodies of free water do not and cannot exist on Mars as it is at present, there is surface evidence that significant quantities of free water did flow in river-beds at one time. What more do you need for life? Yet the Viking Mars-landers found no evidence for the existence of life on Mars. Why not? Is it possible that our current story about the development of life from non-living material, by normal chemical processes and wherever conditions are suitable, is somehow not correct?

One way of solving the problem would be to discover evidence of life on (so far hypothetical) planetary systems about other stars. How could one do this? By picking up radio signals that would indicate transmission by an intelligent agent. This has been the task of Communication with Extraterrestrial Intelligence (CETI). It is a serious and important project; its success would be of great philosophical and practical significance. In order to discuss the relationship between the enormous costs of such a program, and the possibility of success, a formula was devised to encapsulate the problems that need investigation. The number of potentially communicable extraterrestrial intelligent civilizations is given by:

$$R^*f(p)^*n(e)^*f(l)^*f(i)^*f(c)^*L$$

where R is the rate of star formation, $f(p)$ is the fraction of stars which have planetary systems, $n(e)$ is the average number of terrestrial-type planets per system, $f(l)$ is the fraction of those planets upon which life actually develops, $f(i)$ is the fraction of those "living planets" on which intelligent life develops, $f(c)$ is the fraction of those "intelligent planets" which develop the skill of communicating by radio, and L

represents the average lifetime of such civilizations in the radio-communicable phase.

As a summary of the factors that need to be known, this so-called Drake formula is perfectly valid. As a practical means of estimating the chance of success of a listening-in process, the formula is useless at the present time. The only factor in the formula which is reasonably well known is R. Of L, nothing is known. It must be remembered that this is the lifetime of a radio-communicating civilization. Such a lifetime might be limited because shortly after the discovery of radio, nuclear power is discovered and the civilization blows itself up. L would then be negligibly small, because a length of time of the order of 100 years (the period of practical use of radio for communication on Earth) is negligible compared with the actual past and potential future duration of life on Earth of several thousand million years. Indeed, one might expect (hope?) that L would rather be limited by the fact that radio as a means of communication rapidly becomes obsolete. We may be deaf to the chatter of galactic commerce simply beause we are so primitive that we can only think of using radio waves for communication! Of the f-factors we know nothing, save that they are all less than 1 (by definition), and none of them is identically zero (since we exist).

Since a typical f-factor lies between 0 and 1, some authors have suggested that a mean value of 0.5 might be taken. This is not reasonable, however, for it assumes that the unknown f is of the order of 1, whereas it might be 0.01 or 0.0000001. It is the order of magnitude of the fs that are unknown, and it is the harmonic mean of the limits, not the arithmetic mean, that should be taken in the absence of further information. In other words, the most likely value for one of the f-factors, if its order-of-magnitude is unknown, is close to zero.

There remains the question of how one would recognize the output of an "intelligent" civilization. When pulsars, which give out bursts of radio radiation at very regular intervals, were first discovered and their nature not yet known, it was seriously suggested that they might be signals from intelligent civilizations. But a regular beat would surely be the sign, not of intelligence but the reverse. A pendulum or an oscillating neutron star is not intelligent. On the other hand, if all the radio emissions of our own civilization were to be picked up elsewhere, the signals would seem like random noise; and we detect plenty of random noise from the universe. If we wished to communicate our existence and intelligence to other civilizations, what message would we send? A demonstration of the theorem of Pythagoras? A sequence of prime numbers? ...

In fact, the problem of recognizing signals from an intelligent extraterrestrial civilization is very similar to the problem that is the subject of this symposium. Both are concerned with the recognition of "design" in nature, whether localized on planets or disseminated throughout the cosmos, and both are dogged by the same pitfalls and the same ambiguity.

Epilegomena

The ambiguity resides in the question – how far is our description of the external world objective, and how far is it determined by our psychology? We truly see through a glass darkly, and in so far as we share to some degree the same sensual and philosophical filters, that which seems objective (that is, common to all observers) may be in part properties of the filters, not of that which is apprehended through them. Eddington once likened our situation to finding a footprint on the sand, measuring the footprint to find out what sort of creature had passed that way – and finding that the creature was ourselves!

Once we adopt the criterion of "improbability" as a measure of "order," and adopt "order" as a sign of "design," we meet an ambiguity in the meaning of "probability." I have a die, and before I throw it, I ask "what is the probability that I will throw a three on my first throw?" On the basis of my experience of other dice, I might be tempted to say "one-sixth." However, I might be aware of the fact that I had painted an extra dot on the two side, so that there were two threes and no twos. I would then say that the probability of throwing a three was one third. The difference arises in a difference in my knowledge, or state of mind. This is the so-called "prior probability." I also could throw the die a large number of times and discover the bias of the die through the relative frequencies of occurrence of the various numbers; this is probability as a frequency distribution. When dealing with the question of design in the universe, or recognizing "intelligence" on other planets, we are dealing with prior probabilities (since there is only one universe, and only one known inhabited planetary system). So any evidence for "order" in the universe will depend upon our state of mind.

This limitation has come up several times in the course of this discussion. In the first place, in asking whether the biogenic zone of Figure 1 was small, smallness had to be related to the overall size of the diagram (that is, the range of values of G that were considered), and this was limited only by the flexibility of my imagination! Does life depend upon the presence of water and an atmosphere? How flexible is my biological imagination?

In the second place, the prior probability that I assign to a situation will depend upon the theoretical framework within which I imbed my experience. For example, mass enters into Newtonian dynamics in two forms: as gravitational mass, which measures the contribution that a body makes to the gravitational force between it and another body; and as inertial mass, which measures the resistance of the body to having its state of motion changed. Now measurements have shown that the inertial mass is equal to the gravitational mass to at least one part in a million million. Surely this is highly improbable? However, I might hazard a theory that inertia is in fact a result of the gravitational pull of the rest of the universe. Such a theory can be formulated, and it yields a relationship between G, the mean density of the universe, and the age of the universe which is in agreement with observation. On this theory, the equality of the inertial and gravitational masses far from being "improbable," is absolutely necessary because, according to the theory, the inertial mass and the gravitational mass are the same thing. Now for a variety of reasons this theory is not acceptable; Einstein's Theory of Gravitation identifies inertial and gravitational mass at a more profound and more convincing level. But all scientific theories are provisional; it is precisely because my theory is unacceptable that it illustrates my point–that the improbability or otherwise of a "numerical coincidence" depends upon the state of my (theoretical) mind.

However, the observer's state of mind enters even more fundamentally into our problem. Consider again the paradigmatic experiment of the balls in the box. The balls appear all black, and roughly uniformly distributed throughout the box – a "probable" state. However, we have actually been looking through a blue-green filter. If the filter is removed, we find that all the balls in the left-hand side of the box are black, and all the balls in the right-hand side of the box are red – a highly ordered state. The degree of "order" of the atoms in the box depends upon which parameters of the system are considered; if colour is ignored, the system is disordered, whereas if colours are considered, the system is ordered. Given the properties to be included in the definition of "order," it is possible unambiguously to calculate the order of a state of the system; but what determines which properties are to be so considered?

This example of colour may seem contrived; but in a system of atoms in a box, for example, the system proceeds from a state of order to a state of disorder if only positions of the molecules are considered, but proceeds from a state of disorder to a state of order if only the velocities of the particles are considered. Indeed, in so far as an isolated system obeys Newtonian mechanics, every past state and every future state is implied by the complete specification of the present state; any apparent changes of order are then a consequence of our ignorance of the full specification of the initial state, or else our ignoring of certain parameters in our definition of "order."

Thus we find that, since we are condemned to using only prior probability in our search for "order" in the universe, any order that we find is at least in part a reflection of our own state of mind. Indeed, the irony is that in changing from "design" to "order," we seemed to be changing from an aesthetic to an objective criterion; but this was an illusion. In the event, "order" seems to be as subjective as "design." In *The Nature of the Physical World* Eddington wrote: "Suppose that we were asked to arrange the following in two categories: – distance, mass, electric force, entropy [disorder], beauty, melody – I think that there are the strongest grounds for placing entropy alongside beauty and melody and not with the first three." I can now relax, and see "evidence for design" in the wispful symmetry of the Rosette Nebula, without being ashamed of my poetic subjectivity!

Was the sixteenth century wiser than the twentieth in this regard? I look at a painting – say Jackson Pollock *Number 6* – and ask if it shows evidence of design. In this case, the answer is not quite obvious; but I would not try to find the answer by making a chemical analysis of the pigments, to see if their composition was "improbable."

Must we not then say, with Vico, *verum et factum convertuntur*? To understand something, the knower himself must have made it. In seeking evidence for design in the universe, are we simply holding a mirror up to nature in a pointless exercise in narcissism? I think not; but perhaps this is a question that should be addressed in a symposium on "Evidence for Design in the Universe."

Appendix 1 Condensation Formula

Mass of condensation = M	Density of condensation = ρ
Radius of condensation = R	Temperature of condensation = $T°K$
Escape velocity = v_e	r.m.s. thermal speed = v
Mass of hydrogen atom = m_H	Mean molecular weight = μ
Boltzmann's constant = k	Constant of gravitation = G

$$M^4 \tfrac{4}{3}\pi R^3 \rho$$

$$v_e^2 = (2GM)/R$$

$$v^2 = (3kT)/(\mu m_H)$$

$$v < \alpha v_e$$

Hence $M > [81/32\pi\alpha^6]^{1/2}\,[kT/\mu G m_H]^{3/2}[1/\rho]^{1/2}$

James' Criterion for condensation $\alpha = 0.63$

Appendix 2 Largest Spheres

Energy/mass assocated wtih resisting force = η
Gravitational self-energy of sphere $\sim GM^2/R = E$
For stable sphere, $\eta > E/M$
Maximum density of sphere = $\rho_x \sim ml^{-3}$

where m, l are mass, length associated with resisting force.

Hence $R = [3M/4\pi\,\rho_x]^{1/3}$

and $M < [\eta/G]^{3/2}[3/4\pi\rho_x]^{1/2}$

e.g. for water ice, $\rho_x \sim 1$ gm/cc
$\qquad\qquad \eta$ = latent heat of melting = 80 cal/gm

Hence $M < 6 \times 10^{23}$ gm, $R < 535$ km.

Appendix 3 Hypothetical Variation of G

The formulae are not equations; they show the functional variation of M, the mass of a star, with G, the constant of gravitation. The constants A, B, C, and D are functions only of natural constants such as π, and atomic and nuclear constants. It is not necessary to know the values of these constants, if the functions are presented (as in Fig. 1) with G in terms of G_0, the actual value of G, and M in terms of $M_\odot$, the mass of the Sun.

$\qquad GM^{2/3} > A$ $\qquad\qquad\qquad\qquad$ Thermonuclear reactions
$\qquad GM^{2/3} < B$ $\qquad\qquad\qquad\qquad$ Radiation pressure

and for life

$\qquad CL_\odot/10 < G^4 M^{7/2} < 10 \times CL_\odot$ $\qquad$ Luminosity of star
$\qquad G^{-4}M^{-5/2} > Dt_\odot/3$ $\qquad\qquad\qquad$ Main-sequence lifetime of star

W.G. UNRUH

Is the
Physical Universe Natural?

The title of this conference "The Origin and Evolution of the Universe: Evidence for Design?" asks whether or not our present theories about the origin of the universe, the earth, and life itself have any relevance to the old argument from design for the existence of God. That argument is basically that the complex structures and the order which we see around us could not have happened by chance, could not be "natural," but must have been specially created. The contrary argument is that the structures and order are the result of completely natural processes, and can be explained as the outcome of certain simple laws of nature. In the most extreme version of this contrary viewpoint, the universe and we are the unique consequence of natural physical laws. In particular, for the argument from design to have any force, one must show that there is no reasonable way in which the universe could have been naturally produced.

In this paper I shall present some of the approaches that physicists have taken recently in explaining the birth and origin of the universe. In this context, the universe has many features which at first look peculiar, which seem to be so special, when compared with all possibilities one could naively contemplate, that one would be tempted to point to them as evidence of special creation. I will outline some of the approaches which cosmologists and physicists have taken recently to show the converse, that these peculiar features could have been the result of completely natural processes. I will admit that many of these ideas are very speculative, and that in their present form, they do not quite work in explaining all the details one would want them to explain. What is important, however, is that such a natural explanation does seem to be possible. As such, it robs much of the force of any use of these peculiarities in arguing for the necessity of a special creation.

It is ironically fitting that we are holding this conference at McGill, the brief home of Ernest Rutherford, who is reported to have said: "If someone in my laboratory begins to talk of the Universe, I tell him that it is time to leave." This remark illustrates quite graphically the change that has occurred in physics in this century. From being a subject that no self-respecting physicist would discuss as science, it has become the hot topic of the day.

The attitude of Rutherford is understandable because of the severe problems people encountered when they attempted to understand the universe in physical terms. For example, in trying to explain the source of heat and light from the sun by the most energetic processes known – e.g., the burning of coal say – led to absurdly short times (about 30,000 years) for the possible age of the sun. If this was the energy source of the sun, it, and presumably the rest of the universe as well, must have been created very recently, with about the same structure that it has now. On the other hand, the anti-religious prejudice of late nineteenth-century science demanded that creation never took place, that the universe must have existed forever, presumably in more or less its present form. This led to other difficulties. If stars had lasted forever, why had they not heated up the whole universe to the same temperature as that of the stars? Cosmic physics indeed would have to be much different from terrestial physics to allow the stars to burn for such a long time. In either case, the origin of the universe seemed to be a topic not amenable to rational scientific discussion.

The advent of Einstein's theory of gravitation in 1915 broke this impasse. He quickly realized that this theory, in its original form, demanded that the the universe be dynamic, that a universe which remained constant forever could not be allowed. Finding this conclusion repugnant, he fudged the theory to allow such a static universe. It was the observations of Hubble ten years later which showed that the universe was indeed dynamic, just as the original theory had predicted.

Not only did the theory predict a dynamic universe, but it also predicted that the universe could not have lived forever, that it must have had a finite lifetime, which modern improvements on Hubble's observations put at roughly 10 billion years.

It is indicative of the prejudice against creation of the universe, that for quite a while another fudge of Einstein's theory, the Steady State theory, attracted a number of proponents. This theory allows for the expansion of the universe (as observed), but assumes that nevertheless the universe has lived forever in roughly the same form as it has now. To compensate for the dilution of everything which the expansion causes, its proponents hypothesized a continuous creation of matter. The cosmos was thus again different from the earth, both in being eternal, and in obeying different laws of physics from those on the earth. The observation of the three-degree radiation, mentioned by Peebles, and which is believed to be a relic from the birth of the universe, put an end to that theory. The Universe became the same as the earth, suffering birth, change, and perhaps even death just as we do. In the process of removing the heavens from the abode of the gods, where the processes differ in kind from the mundane processes around us, Einstein's theory also raised the possibility that the universe was actually explicable, that one could have a theory of the origin and development of the universe as of any other physical system.

In the standard model for the development of the universe worked out in the late 1950s and 60s, the universe began life about 10 billion years ago, with the matter we presently see in the universe compressed into a region of the order of 10^{-4} centimeters across. The matter at that time was believed to be intensely hot. As the universe expanded, the matter cooled. To serve as a reference for the rest of the paper, Table 1 gives a list of the time after the origin, the size of the region containing all of the matter of our presently observable universe, the temperature, and the density of matter. The times correspond roughly to the earliest time for which the theory makes any sense, the time when the interesting events to be described later occur, the time when helium was formed, the time when there is more matter than radiation, the time when the three-degree radiation we now see was formed and decoupled from the matter, and finally now. This table will be of use as a reference later in this paper.

Since Einstein's theory of gravity plays such a central role in modern cosmology, I should like to spend some time it describing. The description will be simplified in that only those aspects necessary to cosmology will be mentioned. Einstein's key realization was that gravity could be explained by a theory of distances, a theory in which distances change and become dynamic in their own right. One usually thinks of distance between two bodies changing because the bodies move. In fact, it is at first very difficult to think of changes in distance as separate from the motion of the bodies primarily because we often define motion in terms of changes in distance. For Einstein's view of

W.G. Unruh

Table 1

Time	Size of Universe (cm)	Temperature (°K)	Density (gm/cm³)
10^{-44} sec	10^{-4}	10^{33}	10^{94}
10^{-35} sec	10	10^{28}	10^{74}
10 sec	10^{19}	10^{9}	10
10^{11} sec	10^{24}	10^{5}	10^{-17}
10^{6} yr	10^{25}	10^{3}	10^{-20}
10^{10} yr	10^{28}	3	10^{-29}

gravity to make sense, one must define motion separately from distance. Newton's first law of motion allows us to do so. According to Newton's first law, a body at rest will stay at rest; it will not move, unless it is acted on by an external force: no force, no motion. The distance between two bodies will thus change in itself, and not because of the motion of the two bodies, if neither of the bodies experiences an external force; and yet the distance between them changes.

One could of course explain this change in distance by postulating a force which no body could feel or experience except in that it changed the distances between bodies. This is essentially what Newton did in his theory of gravity. It was Einstein's genius to realize that this fictitious force was not needed, that the whole theory could be recast in a conceptually much simpler form if one talked directly about distances changing on their own. (This completely general rewriting of the theory of gravity could only be done after Special Relativity had recast the notion of time into that of a generalized distance in four-dimensional space-time. However, this complication is irrelevant for our purpose in discussing cosmology.)

Thus, in the language of Einstein's theory, the universe expands, not because the matter in the universe is moving further and further apart, but because the distance, say between any two galaxies, is increasing of its own accord. The galaxies are not moving in a pre-existent space but space is continually being created between the galaxies. When the universe was born, there was no space. All of the space we see around us came into being after the universe was formed. In the beginning, all the matter and energy of the universe was crowded into an inconceivably tiny amount of space, so that 10^{-44} seconds after the birth of the universe, everything we now see occupied only about the volume of a speck of cigarette smoke.

Why did the universe come into being? Why did some tiny amount of space come into existence with such an inconceivable density of matter? Neither I nor anyone else can yet answer these questions. The theory breaks down here. All we can say is that the universe must have come into existence, since we are here.

But once we have accepted this existence, we can ask why the universe has the structure it has. The first successful predictions of this theory of the origin of the universe were that of the helium to hydrogen ratio in the universe (about 1 to 3 by weight), and the presence of radiation with a temperature of about 3° Kelvin, as already mentioned. It was the success of these two predictions which established the above picture of the origin of the universe as the standard, known as the Big Bang.[1]

Although this model became standard, it also left a number of unsolved problems. In spite of much effort over a twenty-year period, these problems seemed to have no possible solution. One could well ask whether the only solution was a very special creation. I will look at two of the key problems: first, why does the type of matter we see around us exist at all? and second, why is the universe so extremely smooth and homogeneous on the large scale but clumpy in just the right degree on the mid-scale (i.e., why do stars, galaxies, etc, exist, and are not all clumped so badly together as to form black holes)? It is these problems which have received at least the outline of a solution in the past seven or eight years, and which I wish to present here.

A crucial component (for us) of the present universe is the existence of a type of matter called baryonic matter namely protons. One can divide all possible kinds of matter or energy into two types, called by physicists bosonic and fermionic types of matter. The former is exemplified by radio waves. It is a type of matter (or energy if you will) that is easily created or destroyed. It has no real permanence, and moreover is highly interpenetrable with other matter of the same type. It is extremely difficult to form any stable or permanent structure from this type of matter. A universe made up only of bosons (say light, radio waves, or other electromagnetic radiation) would be extremely different from the present universe. Since all bosonic matter tends to decay rapidly to electromagnetic radiation (or maybe gravitational radiation, which behaves in much the same way), such a universe would not form structures like stars, planets, people, etc.

It is the other type of matter, namely fermionic matter, is needed to form the structures we see around us. Fermions, of which the key examples are protons and electrons, have the properties of matter that we are taught in elementary school: namely, that matter has extension, has at least relative permanance, and has the property of impenetrability. Light rays or radio waves have no difficulty in going right through each other. It is due to the presence of fermions in the universe that we and our world exist. Now, if one measures the number of protons (which equals the number of electrons) and the number of bosons (light quanta or photons) in the universe, one comes up with the ratio of about one to a billion. Although this may seem like an extremely small ratio, it is in fact surprising is that it is not much smaller. Why is all the matter in the universe not in photons? For each type of fermion, like the proton, there exists another type of fermion, its antiparticle. If a particle and an antiparticle meet, they annihilate each other, leaving only electromagnetic radiation. Why did the universe elect to produce so many more protons in the initial stages than antiprotons? Why did it not produce an equal number of protons and antiprotons, since all their properties are almost completely indistinguishable? Now indistinguishability implies that in any system in thermal equlibrium there should always exist an equal number of particles and antiparticles. Since one of the hypotheses of the Big Bang model is that at early time the universe was roughly in thermal equilibrium, the universe should have started out with an equal number of fermions and antifermions which should then have annihilated each other to leave behind a universe with extremely little "useful" matter in it.

Numerous attempts have been made to get around this problem. According to one hypothesis, there perhaps were small statistical irregularities in the very early universe, which produced small regions of particles near other small regions of antiparticles. At the boundary between the regions, the particles and antiparticles annihilated each other, producing intense amounts of radiation, which then pushed the regions apart. It was hoped that this process could eventually produce large regions with a preponderance of matter and others of antimatter. The problem was that detailed calculations could not support this model, and the search for regions where a small overlap of matter and antimatter, and thus the generation of annihilation radiation, could be found have proved futile.

In the past six or seven years, a "natural" explanation for the proton-antiproton abundance asymmetry has emerged. In attempts to understand the behaviour of high-energy accelerator experiments, theories were developed that resulted in very small differences in the detailed behaviour of particles and antiparticles. Although the general behaviour must be the same, so that in thermal eqilibrium both must be present in equal numbers, and so that the total probability for decay or creation of particle and antiparticle must be the same, the precise details could differ slightly. Because of the rapid expansion of the universe, the matter was not in precise thermal equilibrium. Some types of matter could cool slightly faster than other types, and since cooling takes place so fast, the two types may not have a chance to interact and bring each other to the same temperature. If the cooler type of matter interacted to form antiprotons say, and the hotter to form protons, the rate at which the two were formed could be slightly different, leaving slightly more protons than antiprotons. Detailed calculations in these theories of the elementary particles suggested that exactly such could occur at times of about 10^{-35} seconds after the origin of the universe when the temperature was about 10^{28} degrees. If one choses just the right theory, one obtains just the observed ratio of photons to protons.

This was one of the first indications that perhaps some of the fundamental features and structure of the universe crucial for our existence could be explained as the natural outcome of at least some possible laws of physics.

The other set of problems I wish to discuss are those associated with the homogeneity of the universe. As Peebles shows, the universe is amazingly smooth and homogeneous on the large scale. How is this possible?

One answer could be that in the early life of the universe, everything was extremely hot, and close together (the whole visible universe was only a centimetre in size at the earliest moment). Surely such a small hot region would have been very smooth and homogeneous? But this could not have been true, because in those early times the distances were increasing far too rapidly. In order to smooth out any initial inhomogeneities there might have been at the creation of the universe, the different parts would have had to interact with each other. Now interactions can travel at most at the speed of light. Consider a region one centimetre across very early in the life of the universe. In the 10^{-10} seconds it would take a ray of light to travel one centimetre, that one centimetre would have expanded to 10^{20} cm. By the time the light has traveled for the 10^{10} seconds required to go that distance, the universe has increased in size to 10^{28} cm. In fact, only 10 billion years after the origin of the universe could light have travelled from the matter at one edge of the observable universe to matter at the other. Thus the various parts of the universe could not have interacted with each other in any way to ensure that everything was the same in all directions.

Is there any way around this dilemma? Yes, and the solution is something called "inflation."[2] Consider a time of the order of 10^{-35} seconds after the origin of the universe. At that time, only those regions with a size of less than about 10^{-29} cm could have interacted sufficiently to have smoothed themselves out. But one needs a region of about 10 cm at that time. One can imagine expanding the 10^{-29} cm to 10 cm, but at the expense of diluting the energy and matter content of that 10^{-29} cm region. If the matter obeys a very particular set of conditions, one can expand that tiny region while keeping the energy density constant throughout the expansion. This condition is that the matter have a large, negative pressure, a pressure equal to the negative of the energy density. Negative pressures are actually tensions; matter in the region is pulling in on itself very strongly. As the universe expands, it has to do work against this inward pulling of the matter, thus increasing the energy inside the region. In particular, this extra energy fed into the region by the expansion exactly compensates for the energy dilution one would expect as the universe expanded. Furthermore, it turns out that if the pressure obeys this relation, the expansion of the universe takes place extremely rapidly. The universe can expand by the necessary factor of 10^{30} times in 10^{-38} seconds. Finally, matter with a negative pressure is extremely unstable. Like a punctured balloon (within the walls of which a negative pressure or tension exists), such a state of matter wants to collapse in on itself and convert itself to ordinary positive pressure type matter very rapidly. This is fortunate, as the rapid expansion that takes place while the pressure is negative had better not go on for too long, for we know that the size of the universe now is not doubling every 10^{-38} seconds.

Thus, if matter at some high temperature can briefly enter a phase in which such large negative pressures exist, then one could solve the puzzle of why the universe we see is as homogeneous as it is. Fortunately, elementary particle physicists have developed theories for other purposes in which matter can be expected to do just that.

The revised scenario we now have for the developement of the universe goes as follows: in the beginning the universe was created randomly, with no coherence from one region to the next. As the universe evolved and expanded, some very small regions had a chance to interact and smooth themselves out. Then suddenly, when the temperature had dropped to about 10^{28} degrees, matter entered a temporary phase where the pressure turned into an extremely high tension. The universe then suddenly expanded by a huge amount so that one of these small smooth regions grew by a factor of 10^{30} while maintaining the density of matter and energy everywhere. Because of the instability of matter in this regime, this inflationary phase, as it is called, lasted only a very brief time, the pressure again became positive, and the universe resumed its more normal expansion. Shortly thereafter, protons came into being. (Such proton creation must follow inflation, otherwise the density of protons may be too seriously diluted in the inflationary phase.) The whole region of the universe which will develop into the region visible to us is now smooth with the correct number density of protons.

This inflationary phase avoids an obvious difficulty. If the universe had been left with no inhomogeneities on any scale, if it had been completely smoothed out, one would have been left with the problem of where the clumpiness in the matter which we now see as galaxies, stars, and so on came from. Another prediction of the inflationary model, however, is that things are not left absolutely smooth. Because of the rapid expansion during inflation, small fluctuations in the matter density are created by quantum processes, fluctuations that are of the right form to produce the present clumpiness we see. The success of the theory hinges on having the right model for the behaviour of matter at the extremely high temperatures that exist at the beginnings of the universe. This solution to the problem of why the universe is so smooth has become so popular and compelling that the ability to account for a period of inflation is used as a criterion for judging the acceptability of theories of elementary particles.

What I hope I have shown is that it is extremely dangerous to use the special structures that we see around us in the universe as an argument for the existence of God. Many of them seem to be explicable as the completely natural outcomes of natural laws. On the other hand, one can generalize the argument from design to say that it is not the special structures *per se* that show evidence for design, but that the evidence is in the rather special laws which the universe must and does obey in order to lead naturally and inevitably to our universe. God operates not by choosing special conditions for us, or by suspending the behaviour of the natural laws, but by choosing that the universe obey just those laws that make us inevitable. The difficulty with this approach is that we do not know how arbitrary the choice of the laws of nature is. It is possible that there exists only one logically consistent set of laws which also allows the existence of structures sufficiently complex to have self-consciousness. The laws of nature would then have the same force as the laws of mathematics have. There has for example been a long hope in one small subset of the laws, namely the laws of electrodynamics, that the charge on the electron is not a free parameter of the theory, but is forced to have its observed value if the theory is to be self-consistent. If this rather speculative hypothesis is true, the universe is the way it is because that is the only logical way it could be. However, there exists at present no evidence to support this hypothesis, and I had better cease my speculation.

I would like to conclude with a somewhat contradictory thought. We should never forget that a rational analysis of the world is a human endeavour. It is we who have decided that a logical description of the world is of importance; however, it is not necessarily the only possible way of looking at the world. I will leave you with a quotation from William Blake's "A Vision of the Last Judgement": "When the Sun rises do you not see a round Disk of fire somewhat like a Guinea O no no I see an Innumerable company of the Heavenly host crying Holy Holy Holy is the Lord God Almighty . . ."

NOTES

1 A detailed and clear analysis of the Big Bang model is given in P.J.E. Peebles, *Physical Cosmology* (Princeton: Princeton University Press, 1971).
2 On the theory of inflation see A. Guth, and P. Steinhardt, "The Inflationary Universe" in *Scientific American* (May 1984): 116.

IAN HACKING

Coincidences: Mundane and Cosmological

Probability ideas continue to be used in arguments from design. This paper examines the logic of such arguments. (1) Eighteenth-century cosmological arguments are used to illustrate this logic. (2) Difficulties in applying this logic are illustrated by adapting the famous case of Paley's watch. (3) These difficulties are transferred to today's discussions of the "fine tuning" of the universe. (4) The anthropic principle and the idea of many universes have been used to undercut design arguments. It is shown that in connection with Carter's ensemble of coexisting universes, this approach is sound. When applied to a sequence of successive universes, there is a fallacy in probability reasoning.

Are the coincidences found in modern physics so great that we cannot suppose that our universe attained its present state by chance? There are two implied questions, one of physics, and one of logic. The physics is much more interesting than the logic, and it is a pity to turn from the brilliant expositions of Peebles, Unruh, and Ovenden to pedestrian matters of reason. But turn we must. Psychologists such as Amos Tversky and his collaborators have confirmed what all of us have long suspected: at no time do we run greater risk of foolish and elementary error than when we use probabilities.[1]

1 Old Coincidences

There are two fundamental ways in which to think about probability. In one, a probability is part of a way of representing belief. This idea is often called Bayesian, after the Rev. Thomas Bayes, F.R.S., whose work was posthumously published in the *Philosophical Transactions* for 1763. I shall use that approach only once, in section 4. I shall concern myself instead with probability ideas connected with relative frequencies. Here a useful exemplar is a paper published in the *Philosophical Transactions* just four years after Bayes's, by the Rev. John Michell, F.R.S. His title was "An Inquiry into the probable Parallax, and Magnitude of the fixed Stars, from the Quantity of Light which they afford us, and the particular Circumstances of their Situation."[2]

Michell thinks that there are too many apparently double stars. He considers, for example, that there are 230 stars equal in brightness to the pair Beta/Capricorni. He asks, if we were to distribute 230 points at random on a sphere, what would be the probability of getting two points at least as close together as that pair? He finds the odds against that to be 80 to 1. Since there are many known twin stars, the odds multiply.

He is also concerned with larger clumpings of stars. He assumes that there are 1500 stars of magnitude comparable to the members of Pleiades, a group of six stars. "We shall find the odds to be near 500,000 to 1, that no fixed stars, out of that number, scattered at random, in the whole heavens, would be within so small a distance from each other, as the Pleiades are."

Given these odds, he thinks there must be a cause of the clusterings of stars, whether it be due to "their mutual gravitation, or to some other law or appointment of the Creator." He thinks we can conclude "that such double stars, &c., as appear to consist of two or more stars placed very near each together" are "under the influence of some general law." Modern astronomy teaches that Michell was importantly right.

Michell's calculations were challenged over the years, but it appears that he made a legitimate test of significance. Sir Ronald Fisher, founder of the general theory of significance testing, defended Michell in that light. Like everyone else, he had some trouble with Michell's model or calculations or both, but employing Michell's idea, he recomputed the odds against six randomly assigned dots (in a population of 1500) being as close on a sphere as the Pleiades are.[3] He obtained 33,000 to 1, which, as he observes, is quite enough to conclude that there is something significant there.

The calculations do not concern us. Like Fisher, I use Michell as an example in methodology. Michell has a *phenomenon*: the distribution of fixed stars. He thinks that there are too many stars close together. He is surprised, and asks: is this coincidental, or is it sufficiently surprising to make us seek a causal explanation? To answer, we need a *standard of comparison*. This leads him to devise a *model*, a chance device that distributes dots on a sphere. He computes that only two times in a million would we get a clumping of dots as close as the Pleiades. Using this standard of comparison, we should be surprised, or so Michell concludes.

The first thing the example teaches is that there is nothing wrong with using frequency models in contemplating the origin of the universe. In reading the literature recently, I have regularly come across expressions of unease about reasoning about the universe in terms of frequencies. People think "there is only one universe, so there are no frequencies." That is completely baseless. Michell was not contemplating an ensemble of heavens: he was sure there was but one. The frequency is not a matter of the relative frequency with which stars are arranged. It is a relative frequency associated with a model, with a standard of comparison. We consider a chance device on which repeated trials are possible, in order to clarify our minds about stars. We simply do not know how to ask the question "Is this a coincidence or something more?" without constructing a standard of comparison.

Next, we must notice that any model is at best only an analogy. There is more than one way to spray dots on a sphere "at random." Some of the historical debates over Michell's problem are over the choice of model.[4] Those are necessarily informal. Once the model is chosen, it is a mathematical matter: how great is the improbability of a postulated event? The choice of the model is not a matter of mathematics but of sensibility.

Finally, we should notice the conclusion of the inference. It is not, for example, "the probability that the twin stars were so placed by design is 80/81" (or whatever computation we have made). The conclusion is always of the form "Either there is a cause, or else, fortuitously, we have something as surprising as an event that happens only once in eighty-one times on a chance device of a certain sort." This is always the form of the conclusion of a test of significance. A hundred years ago, F.Y. Edgeworth analysed the results of the first card-guessing telepathy experiments. He found that indeed the reported events were exceptionally surprising, and concluded, "Such is the evidence which the calculus of probabilities affords as to the existence of an agency other than mere chance. The calculus is silent as to the nature of that agency – whether it is more likely to be vulgar illusion or extraordinary law. That is a question to be decided, not by formulae and figures, but by general philosophy and common sense."[5] Much later Fisher was to add the caution that we can also simply conclude that if there has been no fraud, then we have had a bizarre run of accidents, reminding us of nothing but the fact that weird runs of happenstance do occur.

Although the modern theory of significance testing arose in experimental work (originally in agriculture), there is a respectable early history of the idea in astronomy. In 1734, well before Michell, Daniel Bernoulli successfully competed for a prize set by the Paris Academy: why do the planets move in planes that are roughly parallel (and why do they not all move in the same place)?[6] Bernoulli begins by investigating whether there is a fact there to explain. Is the rough coplanarity surprising? He constructs three different geometric models, each of which gives a different answer. The odds against this happening by chance are, respectively, 17^5 to 1, 13^5 to 1, and 12^6 to 1. The different answers result from different ways of modelling the inclination of orbits by a random device. But the conclusion is robust: on all three models we should be surprised. Therefore, the phenomenon suggests there is something significant here. It is a helpful reminder that we are only constructing models. There is, lacking more specification, no definitively true answer to the question: how unlikely would it be that seven planes should be chosen at random, intersecting at a common point, and with maximum inclination to each other of such and such?

The example also serves as an illustration that there are different ways of tackling the same problem. The coplanarity of the planets was of interest to Laplace, who used an essentially Bayesian analysis. He reached the same conclusion as Bernoulli. It seldom matters which analysis is used. In the present paper I shall retain the significance level approach, because I think that traditional design arguments have been expressed in that way. It is curious to note a reverse connection. Arguably the first precise significance test ever published was John Arbuthnot's 1710 "Argument for divine providence," which was based on the fact that regularly there are more male live births than female ones.[7]

Let me conclude with an example that is not cosmological but is of local interest. Throughout the nineteenth century it has been known that most atomic weights are integral multiples of that of hydrogen. But there are some glaring counterexamples, such as chlorine at 35.5. Sir William Crooks reported in 1888 that this near regularity is too good to be other than surprising. It cannot be an accident. There must be an underlying cause. That cause was discovered here at McGill University in the work that led Rutherford and Soddy to isotopes.[8]

This example allows me to reiterate a point. Crooks had a model in which a set number of real numbers is selected at random, assuming a uniform distribution on the line up to the largest atomic weight then known. It is exceptionally improbable, on such a set-up, that all but a handful of the numbers drawn are, within a narrow margin, integers. Therefore there must be a hidden cause. In employing this model, Crooks did not imagine for a moment that the atomic weights of the elements might be determined on repeated trials. The relative frequency derives from the model, just as Michell's case of clustering stars.

2   Timepieces

The standard example of an argument of design occurs at the start of the Rev. William Paley's *Natural Theology*. It was published in 1802, at the end of Paley's life. There had already been some 120 years of design arguments promulgated by Anglican divines. I do not intend to discuss Paley directly, but only in the context of P.C.W. Davies' *God and the New Physics*, in which Paley's opening paragraph is quoted.[9]

In the chapter titled "Accident or Design?" Davies says that Paley "articulated one of the most powerful arguments for the existence of God," and quotes:

In crossing a heath, suppose I pitched my foot against a *stone*, and were asked how the stone came to be there: I might possibly answer, that, for anything I knew to the contrary, it had lain there for ever; nor would it, perhaps, be very easy to show the absurdity of this answer. But suppose I found a *watch* upon the ground, and it should be inquired how the watch happened to be in that place. I should hardly think of the answer I had given before – that, for anything I knew, the watch might always have been there. Yet why should not this answer serve for the watch as well as for the stone?[10]

"The intricate and delicate organization of the watch," Davies proceeds to remark, "with its components dovetailing accurately, is overwhelming evidence for design."

There is something wrong here. We have overwhelming evidence for design, but *not* from the intricate and delicate organization of the watch. The evidence that someone made this object is *that it is a watch*, and people make watches. Davies retorts that even "someone who had never seen a watch before would conclude that this mechanism was devised by an intelligent person for a purpose." Now if we use "mechanism" in the literal sense of the word, then once again the evidence is not the delicate organization, but *that it is a mechanism*. People, and only people, are mechanics.

Davies speaks of the components of the watch "dovetailing accurately." That is an odd metaphor, for dovetailing is something that you learn to do in carpentry class, not in the machine shop. (Paley, I should insert, is much more careful with his metaphors.) So suppose we have a carpenter from ancient times, Joseph, say. He has never seen a mechanism, but, *per imposibile*, he finds on a desert road to Egypt the watch of my great-great-grandfather which I still use to tell time. I doubt that he would know what to make of it. My "mechanism" would seem to him to be bad carpentry. Joseph would see no "dovetailing." But I suspect that he *would* take my ancestral watch to be a human artifact, not because of the way the odd bits fit together, but because it is worked gold and has the Roman numerals from I to XII on its face. Joseph would have been familiar with these. He would know that only humans work gold and make Roman numerals.

The same would go for Paley if he found a working quartz watch on his heath. There would be no mechanism at all, but he would know it was of human origin because it displays something like Arabic numerals, and these (he and his friends would soon discern) correlate wonderfully with Greenwich Mean Time. Only people produce Arabic numerals and only people represent the passage of time by numbers. Mind you, the LCD watch would be a *marvel*, and it might completely baffle Paley and his peers. But the "overwhelming evidence for design" would have nothing to do with mechanics. The evidence would be that the marvel kept time in a human system of representation.

Unlike Michell's discussion of the twinned stars, Paley's watch teaches nothing about a serious argument for design based upon coincidence. I respect Paley's writing on other grounds (it includes what may be the first detailed conceptualization of self-reproducing automata!) and would give an account of his fallacies quite different from anything appropriate to this volume. Yet we can adapt his story. Suppose that I have found not a watch on the moor, but am walking west along the tree line in Northern Quebec and Ontario. To the left are stunted trees, to the right is tundra. One day, when my old pocket watch reads III, I rest beside a slightly taller tree, which I note is partly ringed by clumps of trees. The more I look, the more I begin to notice something odd. The shadow from the taller tree falls over a clump of three trees. On one side of that clump is a clump of two trees, while on the other is a clump of four, and so on. Have I found a living sundial? Chance or design?

Here we have a real question about design. Had I found the trees thus arranged in a semi-formal garden, I might well bet on design at once. But up here in the tundra? I do know that lonely men in the North do very, very odd things. One of them, fifty years ago, could well have tried to plant a sundial of scrub spruce trees. Design is by no means to be ruled out *a priori*.

One question to address is that of *tolerances*. What would I count as a sundial? Remember that an immobile sundial tells the right time for only small parts of the year, which is why a correction curve is engraved on the better dials. It tells how much to add or subtract from the shadow each month. I also have to decide what class of configurations of trees I would count as a sundial. How spread out, how far away, can the clumps of trees be? Would I count this as a sundial if at the 11 a.m. clump there were only ten trees but a spot where a tree might have been? I do know that there is a good chance that an eleventh tree might well have died. This is a land of poor soil and harsh climate, where trees hardly survive anyway.

In this case we have one advantage over most questions about coincidence. We know, up to the question of tolerances, what we are supposing to have been designed, namely a sundial using canonical representation, except with clumps of n trees representing the numeral n. You would only jeer at the gardener who told you that the herbaceous border was a sundial. "It keeps perfect time; you see, the sorrel casts its shadow now over the basil at 2:27, and that is the time told by the basil. It casts its shadow over the thyme at 10:07 and over the marjoram at 10:20 ..." That is a way to prove that every thicket is a sundial.

Unlike this absurd herb garden, we suppose that our arboreal sundial conforms to our canonical way of representing the day. Let us demand that it is "designed" for only twelve hours of the day, 7 a.m. until 6 p.m (a rather stunted sundial for a northern summer). The "hours" are represented by pie-shaped wedges of trees such that the centre line of the wedge with three trees coincides with 3 p.m., and so forth. Let us be tolerant about how wide the wedges can be. Let us also not factor into our surprise that there are no trees except in the region between 7 a.m. and 6 p.m. – we can readily attribute that to the infertility of this inhospitable terrain. There are seventy-eight-hour trees; we tolerantly conclude that is about as many trees as a plot such as this could grow. Even if we consider all these aspects of our dial as in no need of explanation, our dial looks quite surprising.

One elementary way to model the problem is to think of the hour-wedges as cells – or wedges on a dart board. We imagine a chance device that allots seventy-eight indistinguishable darts to the wedges, with it being equally probable that a dart falls in any wedge as any other. There are 3.65×10^{13} – roughly thirty-six English billions – of such equally likely allotments, only one of which has seven dots in the first wedge, eight in the next, and so forth. Thus in this model, the odds are 3.65×10^{13} to 1 against getting the darts arranged as on our dial.

That sounds impressive: this sundial must be planted or else something truly unusual has occurred! Unfortunately a rough calculation based on several forestry handbooks indicates that there are about 8×10^{13} trees in naturally forested parts of the continent. So I expect that there are a number of natural sundials in the woods, somewhere. It may be surprising that I came upon one and noticed it, but the fact that I have found a "sundial" is no evidence at all of design.

Notice, moreover, that many other patterns might have struck me on my walk. I might have found a clump of trees spelling out my name, I A N. Anyone who has used a stereoscope to study aerial photographs of forests will know that after a while all sorts of patterns leap to the eye. It is altogether likely that if you are alert to that sort of thing, you will find many "human" patterns – letters, dials, pornographic images, etc – if you spend a lot of time in the woods. The situation is completely different if you *specify in advance* the pattern, the tolerances, and the region in which it might be found. Thus suppose we believe that fifty years ago the mad prospector Paul Leclerc found a gold-mine where I am walking, and that he buried his trove and planted a living sundial to mark the spot. I am looking, within five square miles, for a sundial. It is extremely improbable that within a pre-assigned area of that size, there should occur this pre-assigned pattern, a sundial. He who finds a dial in that situation is well entitled to his cry of "Eureka!"

I said that there was nothing in principle wrong with the significance tests of Michell, Bernoulli, or Crooks. Why do they not succumb to my criticisms? They did not specify the phenomena – clustering, coplanarity, or being integral valued – before the observations that they used.

One obvious difference is that we do not have a population of 500,000 (or as Fisher has it 33,000), heavens, with which we are or might be acquainted. Bernoulli did not know of 13^5 planetary systems, nor Crooks of a manifold of sets of elements.

It may be protested that we know a lot of other physical systems that are formally analogous to these. If we contemplate a large population of such, may we not reduce Michell's clusterings to coincidence? There are lots of near spheres with dots on them (dirty oranges, for example). Why do we not factor these into our calculations? It is surely because we have a crude concept of homogeneous populations – homogeneous with respect to possible causes that might act on them. Whatever makes an orange dirty has nothing to do with how the stars get in the sky.

There are troubling cases. For example, suppose I find that the trees in an oasis in the Sahara form a sundial? I would certainly vote for design. Why do I not lump together Canadian trees and those on the Sahara? In the first instance, because I know that most oases have been tended by humans for millennia. But even if I did not know that, I should be strongly inclined to regard the Saharan palms and Canadian spruce as inhomogeneous populations. Why? In a short compass I can do little more than repeat Edgeworth's phrase, "general philosophy and common sense."

In his chapter, "Accident or Design?" Paul Davies considers someone picking up a pebble on the beach:

If the selected pebble had turned out, for example, to be exactly spherical, surprise would indeed have been justified, even if its spherical nature had not been specified in advance. A sphere is a very special sort of shape with the property that is highly regular. Even after the event the random selection of an exactly spherical pebble would be regarded as an event deserving some sort of explanation. (170)

I disagree for many reasons. Note the question of tolerances. "Exact" spheres of this size do not exist in nature nor can they be made by people, or so I am informed by the experimenter C.W.F. Everitt, whose group spent several years trying to make spheres the size of a ping pong ball for use in a small gyroscope in a satellite as part of a test of the general theory of relativity. So Davies' "exact" means "within tolerances."

Note the question of "random." Any student of probability knows it is very hard to pick things at random. I am told by the statistician P. Diaconis that when cohorts for the American draft were selected by pretty girls drawing marked balls from a bag "at random," the draws, after the first few draws, were unintentionally highly regular. As for picking up objects at the beach – unwittingly one reaches for an "interesting" object.

Finally, consider the "specification in advance." One of the joys of going to the beach with children is that they are forever picking up interesting objects. One is bound to pick up something "highly unusual" walking on the beach. I do not deny that one may be curious about the pebble, or about the sundial. But both should be looked upon as mere coincidences until more information is to hand.

3   Fine Tuning

Cosmological reflection, in recent years, has been much aided by noticing a lot of curious coincidences. In order to get from the big bang to us, there would have to be very stringent constraints on what might otherwise be regarded as possible initial conditions, and possible values of parameters – fundamental constants – in physical laws. If the initial conditions were a little bit off, or the parameters were just a little different from what they are, we would not exist. But we do. Hence, it is urged, the universe must be very "finely tuned." There is a fine adjustment of means to ends, as they used to say in the eighteenth century.

From here we branch in two directions: either we consider whether a designer designed this fine tuning, or we consider whether the anthropic principle makes that consideration otiose. I discuss logical aspects of the anthropic principle in section 4. Here we have to consider whether the fine tuning is surprising without yet introducing anthropism.

In this volume we have papers by students of the physical, geophysical, and life sciences that seriously question the "fine tuning" hypothesis. That is, the existent world is not so surprising as has sometimes been made out. Nevertheless there is a logical question independent of current opinion: what might be inferred from "fine tuning" if it did exist?

This is not a question that can be answered in general. Each alleged bit of tuning requires its own modelling. I hope this is evident from my ramblings about trees and quartz watches. There is no general answer to the question: how surprising is a coincidence? You can only look at individual coincidences and see how to model them.

Take, for example, the matter of initial conditions. It is held to be surprising that in the beginning there were almost as many antiprotons as protons, but not quite as many. This fact is instructive, but surprising? No one knows how the protons and antiprotons got there, so the hired probabilist will say, imagine an incredible multitude of items being tagged as pro and antipro by tossing a fair coin. What you would expect is that about half the time there would be a slight majority of one and about half the time a slight majority of the other. Almost never would you get exactly as many pros as antipros (the case in which everything becomes light). So the chance of a slight majority of pros is about fifty-fifty, and there should be no surprise.

I do not imply that there are no models more interesting than my trivial one. I intend only to remark that one cannot model coincidence in a vacuum, if I may so put it.

After such foolishness we should address the question of "fine tuning" more seriously. The question arises first of all from supposing that there are definite laws of nature that are the framework of all possible universes. These laws contain free parameters that are fixed in our universe. In our universe we call these fixed parameters "fundamental constants." For our universe to arise, there must be a happy combination of just the right initial conditions and of an ordered set of values for the parameters. It is a striking discovery of recent cosmology that this ordered set must be drawn from a minute subset of the class of all possible ordered sets of parameter assignments. It can be argued in some cases that this subset is of measure zero in a suitable topology.

These conclusions are remarkable in themselves, and have been very useful in suggesting further cosmological problems and their solutions. What I am about to say in no way diminishes my respect for this powerful program.

We are here today to ask if these conclusions indicate some impossible coincidence, a true surprise. There are many questions about how to model this problem, but I shall restrict myself to the one that the physicists do not ask: what is this framework of natural law?

Galileo had an excellent answer. God wrote the Book of Nature. He wrote it in the language of mathematics.[11] The sentences in the Book are the laws of nature. If this is how you think of the framework, you will not care much about an argument from design for the existence of God, for you are presupposing that there is a designer, indeed a Writer.

There is a sense in which most physicists act as if they did have that view of laws of nature. It can be argued that acting as if you believed that – reasoning in the Galilean style – has been the key to most progress in the natural sciences. I do not want to knock the Galilean style. But we do have to question it if we are to consider the question of coincidence seriously.

My problem is that I cannot attach *any* clear sense to the idea of a framework of natural law existing "before" the first three minutes, and waiting to have the parameters filled in. In fact I am more sceptical than that. Not only can I not imagine that the laws of all these sciences were "there" at the big bang, but I can hardly imagine that there were laws of the psyche before there was a psyche, or laws of life before there was life, or even the laws of the formation of the planets before there were galaxies.

Let me characterize two pictures of the laws of nature. One is the Galilean picture. On this picture there are preordained laws of nature, and it is the task of science to reveal them. We believe that we have had some success, and that our present conjectured fundamental laws are true laws.

The other picture I shall call Duhemian, after the French physicist, historian, and philosopher of science.[12] Pierre Duhem incidentally was a devout Christian. But he was very sceptical of the absolute reality of laws of nature. He thought that the laws of nature were our representations of a complex world, and that there could well be alternative representations. He thought that laws were ways for us to make sense of phenomena, "to save the phenomena." They are the structures that we impose on experience. Not that we can impose anything: one point of the experimental method is to set out constraints. One point of our representations is to enable us to make predictions and for us to create new phenomena.

Naturally there are many philosophies of science between these two extremes, but they provide us with some access to our problem of coincidence.

If we were to use the Galilean picture, we would need some way to understand the pre-ordained framework. I can understand it only in terms of a designer, and a pretty divine one at that. The task of fixing parameters seems to me absolutely trivial compared to the task of writing those laws. Thus it seems to me that in the Galilean picture, the "fine tuning" argument for design is irrelevant because a designer is presupposed. But I shall return to this point.

What of the Duhemian picture? Here the laws of nature are patterns discovered by us to cover the phenomena. Of course we cannot use any pattern, no more than we can find any pattern in a forest. But the patterns are human representations. More importantly, they are not patterns with free floating parameters that we then determine. When Newton wrote down the law of gravity, the old fox had a remarkably close approximation of G in mind. Finer determinations of G have occupied measurers ever since, but the law of gravity came with G fixed within a narrow band. There was not in the first instance a parameter e occurring in numerous accepted laws, with the e waiting to be filled in by Millikan. On the contrary, it was Millikan's determination of charge that made much of the scientific community believe that there are electrons, and that there is such a constant as e.

The laws that we devise start with approximately fixed parameters, or we would not think of the laws at all. After we have devised the laws, then we can generalize so as to get the general form of the law. But it is not surprising that the parameters of the universe all fall in a narrow subset, because we would not have thought of these representations without precisely these parameters. There is no reason to believe that "observers" on a universe with substantially different parameters would ever have thought of our laws. Thus I propose that the "explanation" for the numerous coincidences lies in our pattern-finding procedures. This is much like the explanation I would give for someone who finds "some" pattern in the trees.

To return to the Galilean picture: that is much more like the case of one who has a set pattern, a sundial, a pattern set in advance (in this case, by God) for which one is looking. *We* do not know the pattern in advance, but there is just *one* pattern there, the one written in the Book of Nature. Just as with the sundial it is possible to do some modelling in which remarkable coincidences occur. But since this makes sense only from within a philosophy that presupposes a designer, the consequences are not exciting.

We can redescribe the difference between Duhemian and Galilean points of view in the following way. For the Galilean, there really are laws with parameters, so it makes sense to speak of a space of all possible universes, namely with all possible values of fundamental constants assigned to the laws of nature. For the Duhemian, such a space is a mere abstract construction, an abstraction from the representation that we at the moment find fitting for the phenomena with which we are acquainted. A seriously different universe, for a Duhemian, could well be imagined to have observers, who would concoct laws for their phenomena. But there is no reason to think that our universe would occur in the space they made by generalizing from our laws, nor any reason to think that their universe would occur in the space we invented.

Notice that the Duhemian does not say that we observe a universe with such-and-such parameters, and that only a universe with those parameters could be observed. That is an anthropic thought of the sort to be discussed below. It makes sense only within a Galilean picture. The Duhemian suspects that the parameters of our representations of the universe fall in narrow bands because we make those representations. The parameters, in approximation, often come before the very laws in which they are embedded. As often as not the laws are made to fit the parameters, and not the other way about.

I am here faithful to Duhem, who was very sceptical about the idea that the constants of nature are unique real numbers (which it is our task to determine). They are part of our system of representation, just like sundials and other patterns in the woods.

4 Many Universes

There are two distinct notions of many universes. (1) There are many coexistent universes, although ours is the only one of which we can have positive knowledge. (2) There are many successive universes but we are able to have positive knowledge only about the most recent, namely the one that we are in.

The physical reasoning behind the two ideas will obviously differ. For examples of notion 2, Boltzmann thought of a sequence of universes separated by large passages of time. Wheeler has proposed an oscillating sequence of universes: each universe finally collapses and is succeeded by another. Both physical ideas are alien to (1), which I shall soon describe.

In philosophical writing, however, there is often a suggestion that from many logical points of view the two ideas are equivalent. Thus in his chapter "Accident or Design?" Davies writes that according to (1):

there *is* an ensemble of universes of which ours is but one member ... [While in (2)] Boltzmann's hypothesis, mentioned above, is *logically identical* to that many-universes theory. His universes occur sequentially, but the organized phases are separated by such enormous chasms of time that they are all but physically independent. (171ff; emphasis added)

I wish to show that from the standpoint of probability logic (1) and (2) are strongly non-identical.

Both ideas have commonly been discused in terms of Brandon Carter's "*anthropic principle*, to the effect that what we can expect to observe must be restricted by the conditions necessary for our presence as observers."[13] Nobody would quarrel with that proposition.

Carter formulates this principle in connection with coexistent universes. It is worth recalling his motivation. He was once taken with "various exotic theories" proposed by Bondi on the basis of "large number coincidences." But then he came to realize that one could predict these very coincidences on the basis of conventional theory, plus the proposition that the known world must have features capable of supporting life such as ours. Similarly, one of Bondi's starting points is that the universe is almost isotropic. Collins and Hawking concluded in a similar vein that "the answer to the question 'why is the universe isotropic?' is 'because we are here.'"[14] In other words, we do not have to draw Bondi's exotic inferences.

Carter also introduced "what may be termed the *'strong' anthropic principle* stating that the universe (and hence the fundamental parameters on which it depends) must be such as to admit the creation of observers within it at some stage." Carter is able to draw further deductions from this assumption. Clearly it, unlike the (weak) anthropic principle, has content; it is, as Carter understates the matter, "rather more questionable."

I have gone on at length to make the following point: thus far there is no mention of more than one universe. Carter turns to that notion in words of exemplary caution that all philosophers who write on the anthropic principle should engrave on their desks:

It is of course always philosophically possible – as a last resort, when no stronger physical argument is possible – to promote a *prediction* based on the strong anthropic principle to the status of an *explanation* by thinking in terms of a world ensemble. By this I mean an ensemble of universes characterized by all conceivable combinations of initial conditions and fundamental constants (the distinction between these concepts, which is not clear cut, being that the former are essential to local and the latter to global features).

There follows much more of comparable clarity. For example, the strong anthropic principle would define a cognizable subset of all possible universes. The cognizable subset might be of measure zero on a suitably defined topology, but "subject to the condition that it is possible to define some sort of fundamental a priori probability measure on the ensemble" we would hope to show that "most" members of the cognizable subset would be features of interest to us.

It has now become common to run together Carter's ideas and the earlier branching-universe idea of Everett's. The branching universe was a proposal to solve the measurement problem in quantum mechanics. Its measure-theoretical features – always of interest to the probabilist – do not evidently correspond to Carter's. It is utterly different in motivation from Carter's cosmology, although it may seem that buying one model of many-universes helps develop a taste for others.

In my opinion Carter's reasoning is impeccable. It is true that at present there exists a furious debate among philosophers of science as to whether "inference to the best explanation" leads to truth.[15] That is, if you find that X is the best available explanation of Y, and know Y but find it surprising, then, lacking other grounds, should you reasonably believe X? Leaving aside my own personal scepticism about inference to the best explanation as a mode of inference, I have no doubt that a world ensemble of many actual coexistent universes is an explanation of the sort intended by Carter. Thus I pass by the question of whether one should believe it because it is an explanation.

I shall make a trivial analogy to Carter's impeccable reasoning. It is to be contrasted to my forthcoming analogy for sequential universes. Suppose that out of the blue on my fiftieth birthday I receive a telephone call from Paraguay. I am informed that I have just won $500. Shortly before my birthday a person I have never heard of (so I am informed by the interpreter on the telephone) has dealt spades, in order, from a short deck of cards, with the understanding that if he or she did this, I was to be telephoned with the good news that I had won the $500 as a birthday present. A short pack consists of ace up to seven of each suit; it has twenty-eight cards. The spades are dealt in order if the first card dealt is ace of spades, the next deuce of spades and so on.

What is all this about, I demand, but my informant is sworn to secrecy, and hangs up. Next week I do indeed get a cheque for $500 in the mail. Notice the crude analogy to (weak) anthropism. I would not have known of this crazy set-up unless I had won.

There are two bizarre features to this tale. One is the arrangement with the person in Paraguay (and I do check with the phone company that the call did emanate from there). The other is the reported deal. The chances of such an outcome from a well shuffled short deck are only barely better than 1 in 6 billion (1:5,967,561,600).

Perhaps the best explanation is that this is an agreeable joke played by a friend travelling in South America. But I can concoct the following explanation, call it E. I don't say E is a good explanation, but only that, if true, it would completely explain my lucky but preposterous telephone call. According to E, virtually everyone in the world but me was given a short deck of cards, and various facilities, such as an interpreter, a dealer if the recipient were a child that was too young, and so forth. But the decks are all stacked. Each of 5,967,561,600 people has been given a different stacked deck. All possible decks are assigned to people "at random." The instructions with each deck are identical: if spades are dealt in order, then I am to be telephoned with the good news. E would perfectly explain my mysterious telephone call. *The explanatory power of E has nothing to do with probability*. It is absolutely certain that I shall come up a winner, exactly once. We can build in a probability over this "cognizable subset," namely the probability that the person who deals, or has dealt for him, "my" deck, is going to phone me, and that the support systems will work. Let us say I think that by and large most people do nice things given the opportunity, so there is a good probability within the (unit) subset of cognizable outcomes that I shall learn of my good fortune. Notice that I have just given a strongly anthropic explanation. It is absurd, but has no logical defects at all.

Now let us turn to the idea that Davies calls the "logically identical" idea of a sequence of universes. This too has been deemed explanatory. Our universe is very improbable, but of course only we could observe a universe we are in. However, in a sufficiently long run of universes, even the most improbable will turn up sooner or later, so our universe should not be a surprise. The assumption of a long previous run of universes helps to explain the existence of our universe. Thus an inference to the best explanation (combined wtih the physics of a Wheeler or a Boltzmann) would make us believe that there were many universes before ours.

There is a fallacy in the above paragraph. The assumption of a long run of (stochastically independent) universes does not help to explain the existence of our universe.

For an analogy, I get a telephone call from the Paraguay Club at the foot of my street at home. The call does not arrive on any designated day, such as my birthday, but does come out of the blue. This time I am told that the Club has a rule that it always shuffles a short deck at the start of play, and deals out the first seven cards. If these are the spades in order, I am to be phoned (I live only a block away) and told I have won $500. We retain the (weak) anthropic character of the example: I would never have known what they do at that seedy establishment had I not "won." I am made a further proposition. "Do you think we just introduced the rule, or have we had it in place for the innumerable times that we have opened play? You get another $500 if you guess right. Hint: there are only two possibilities. This is the first time, or, we have had this rule in place as long as you have been in the neighbourhood."

There are two hypotheses: M for "many deals" and L for "single deal." My further information is A: I have been telephoned today. Notice that my information is not that I have been phoned "at some deal or other" or "at some time or other." Even to use those words would tend to beg the question in favour of M. (Comparably, what we know is that we exist in *this* universe, not some anthropically possible universe or other.)

Undoubtedly most people prefer hypothesis M. Should they? No. Suppose I have an initial assessment of the credibility of L versus M. I may for example find it more incredible that the club would have this inane rule for a long time, than that it should have put in the rule today as a joke. This initial assessment would be reflected by my proper probabilities $P(L)$ and $P(M)$. Now I wish to factor in the observation A, to obtain a posterior probability $P(L/A)$. By Bayes's rule,

$$P(L/A) = \frac{P(L)P(A/L)}{P(L)P(A/L) + P(M)P(A/M)}.$$

$P(A/L)$ and $P(A/M)$ are identical. To suppose otherwise is to commit the so-called gamblers' fallacy. This is the fallacy of a person who sees that on a roulette assumed to be fair, red has come up a dozen times in a row, and so bets black. For, he reasons, black and red are equally probable, and in the "maturity of chances" black has to start catching up with red soon. He reasons with a similar error that if a pair of fair dice have been rolled a hundred times with no snakes' eyes, now is the time to start betting on the otherwise unlikely double-one. This is the fallacy of thinking that fair games have memories. But the probability of double-one is 1/36 after 100 failures just as it is 1/36 at the start of play. In general the probability of getting a rare A after a long run of known As is exactly the same as getting A on the first deal. $P(A/L) = P(A/M)$.

Hence $P(L) = P(L/A)$ and $P(M) = P(M/A)$. I can of course have my views about L and M – views about how night clubs work, represented by $P(L)$ and $P(M)$. But the fact that a very improbable event has occurred should not influence my prior probabilities.

In short: the improbability of A does not increase the probability of there having been a very large number of previous trials. The improbability of our universe does not make it any the more probable that there has been a long sequence of preceding universes. Of course, on other grounds I may think that is likely.

But may there not be a case for inference to the best explanation? We commonly reason like this. I dip my hand into a sack of apples, and draw out, haphazard, one with a worm in it. The event is W. Two explanations are to hand. Most of the apples are wormy: M. Or, this was bad luck, and I got the only wormy apple in the bag: L. Clearly W is far more likely on M than L. We prefer the hypothesis of greater likelihood and, on one view, this has something to do with explanatory power. (We can also reach a preference for M over L on Bayesian grounds.) But unless we have inequalities like that, we have no basis for an inference to the best explanation. As we have seen, there are no such inequalities in the sequential universe case.

The fallacy arises from confusing two propositions. First is the truism that a highly improbable event is far more likely to occur some time in a very long sequence of trials than in a single trial. Second is the falsehood that, given that an improbable event has just occurred, there must have been a long sequence of previous trials. Carter's ensemble of actual coexistent universes has an explanatory power lacked by the assumption that there have already been ever so many successive Wheeler universes. This is a purely logical point that has nothing to do with the credibility of any of these fascinating stories about Everything.[16]

NOTES

1 Amos Tversky and Daniel Kahneman, "Judgement under Uncertainty: Heuristics and Biases," *Science* 185 (1974): 1124; "The Framing of Decision in the Psychology of Choice," *Science* 211 (1981): 453.

2 *Philosophical Transactions* 57 (1767): 234, esp. 243-50.

3 *Statistical Methods and Scientific Inferences* (Edinburgh: Oliver and Boyd, 1956), 39.

4 For references to some of the debates, see Fisher, and Isaac Todhunter, *A History of the Mathematical Theory of Probability* (London: Macmillan, 1865), 332, 393, 491.

5 "The Calculus of Probabilities Applied to Psychical Research," *Proceedings of the Society for Psychical Research* 3 (1885): 199.

6 For a summary, and references to subsequent work on the problem, see Todhunter, *History*, 222, 273, 475, 487, 542.

7 *Philosophical Transactions* 27 (1710): 186.

8 "President's Report," *Journal of the Chemical Society* 53 (1888): 487.

9 *Natural Theology; or, Evidence of the Existence and Attributes of the Deity* (London: R. Faulder, 1802). The passage quoted by Davies occurs on p. 1 of the first edition.

10 *God and the New Physics* (London: Dent, 1983), 164; cf. *The Accidental Universe* (Cambridge: Cambridge University Press, 1982).

11 *Discoveries and Opinions of Galileo*, trans. and ed. Stillman Drake (New York: Doubleday, 1957), 237.

12 *The Aim and Structure of Physical Theory* (1906; New York: Atheneum, 1962). A recent recasting of some Duhemian views is Nancy Cartwright, *How*

the Laws of Physics Lie (Oxford: Clarendon Press, 1983).

13 "Large Number Coincidences and the Anthropic Principles in Cosmology," *Confrontation of Cosmological Theories with Observational Data*, Proceedings of the Second Copernicus Symposium, Cracow, 1973, ed. M.S. Longair (Dordrecht and Boston: D. Reidel, 1974), 291.

14 "Why is the Universe Isotropic?" *Astrophysical Journal* 180 (1973): 334.

15 Sceptics include Bas C. van Fraassen, *The Scientific Image* (Oxford: Clarendon Press, 1980), and Cartwright, *How the Laws of Physics Lie*.

16 For further discussion see my "The Inverse Gambler's Fallacy: The Argument from Design. The Anthropic Principle Applied to Wheeler Universes," appearing in *Mind* 96 (1987).

JEAN-PAUL AUDET

Directionalité, intentionnalité, et analogie dans l'approche de l'univers

J'ai pensé que, le premier à prendre la parole dans ce colloque, il serait opportun que je donne à mon intervention la forme de prolégomènes: distinction et définition de quelques notions fondamentales qui, de mon point de vue à tout le moins, paraissent les plus utiles aux échanges que nous entreprenons aujourd'hui. Directionalité et intentionnalité se retrouveront ainsi au terme de l'élaboration de cet ensemble de notions. C'est pour marquer dès le départ l'orientation fondamentale de ma démarche que je les ai fait paraître dans mon titre. Toujours présentes d'une manière le plus souvent implicite, directionalité et intentionnalité n'occuperont cependant pas à elles seules tout le champ de ma pensée. Complexité oblige!

D'autre part, l'analogie fournira inévitablement, il me semble, l'un des outils conceptuels les plus adaptés à notre propos, quelles que soient la discipline, ou les disciplines, à laquelle, ou auxquelles, chacun de nous voudra se rattacher plus spécifiquement. Mais pour que l'analogie fonctionne un peu efficacement dans un sujet aussi vaste et aussi complexe que le nôtre, il est nécessaire, je pense, que je prépare la voie à son bon usage en introduisant d'abord trois distinctions importantes que nous devons avoir constamment présentes à l'esprit. Je commence donc par présenter successivement ces trois distinctions.

Premièrement: distinction entre monde et univers; intérêt de cette distinction dans le problème qui nous occupe. Biologiquement, on peut dire si l'on veut que l'homme naît dans l'univers, au plan des structures de base et des grandes lois de la réalité et de l'histoire cosmiques. On considère alors l'homme comme une organisation particulière et limitée d'un certain nombre d'atomes: l'organisation la plus surprenante connue à date, la petite "galaxie" pensante que nous constituons, chacun d'entre nous dans l'ensemble de l'univers.

Mais, culturellement, l'homme naît d'abord et principalement dans le monde à la fois restreint et extraordinairement complexe où nous sommes. D'où, il me semble, certaines conditions préalables, et inéluctables, à notre approche de l'univers. C'est une approche seconde, acquise sur le tard au fil de l'histoire de notre espèce, transmise corrélativement par voie de simple apprentissage, d'un petit nombre à un petit nombre. Ce n'est pas l'approche native et originelle à laquelle accède obligatoirement à sa naissance chaque individu normal de l'espèce (par exemple, à travers les images de base de toutes les cultures, et de toute culture, telles les images d'horizontalité et de centralité, qui appartiennent en fait au petit nombre de ce que j'appelle les "images-guides," inscrites génétiquement au plan des "représentations" primaires parmi les éléments indispensables de notre équipement de survie comme individus et comme espèce).

Ainsi, nous *partons* du monde, pour le meilleur et pour le pire. Mais vers quoi? – et sans doute aussi, pourquoi? En d'autres termes, du point de vue d'un sens ultime (recherché mais non possédé), le monde où nous sommes nés, ne nous suffit pourtant pas, ou ne nous suffit plus (la révolution copernicienne, en ce qui concerne l'homme, *habitant* de cette planète, a eu lieu, et a encore lieu, dans la culture elle-même, bien plus que dans l'étude restreinte de la situation de la terre dans l'ensemble immédiat dont elle fait partie).

Reprenons dès lors notre question principale, celle de ce colloque: quel est *l'autre que le monde* vers lequel nous nous dirigeons (intellectuellement ou spirituellement), lorsque par paliers analogiques successifs, nous nous risquons à *décoller* (momentanément) de notre lieu de naissance (sorte d'état d'apesanteur intellectuelle) en vue de jeter un regard curieux au-delà de notre berceau planétaire? Ou encore, réciproquement, *l'autre que le monde* n'est-il en définitive, et en dépit des apparences (immensité de l'univers), qu'un *même que le monde* sous une image à peine reconnaissable par suite de la dispersion et de l'allongement proprement illimités des points et des lignes qui la composent?

Ma deuxième remarque, d'autre part, voudrait toucher brièvement à un problème connexe et plus précis: celui du rapport de la centralité et de l'immensité dans l'ensemble de notre vision analogique du monde et de l'univers.

A ce propos, nous ne faisons plus assez attention, me semble-t-il, au fait que l'immensité (matière-espace-temps) est une acquisition récente de la culture, et notamment de la culture occidentale. Au reste, ce caractère d'acquisition récente n'est pas dû simplement au hasard d'un "génie" qui nous aurait manqué jusque-là au moment propice, ou à la seule lenteur relative du développement de la science. Il appartient au contraire au fond le plus stable (sinon le plus immuable) de notre rapport originel au monde.

Mais comme, au premier regard, nous ne voyons pas beaucoup d'intérêt théorique à explorer pour lui-même ce rapport originel (dépassable en principe, croyons-nous, et peut-être dépassé en fait) nous nous comportons en conséquence à l'endroit de l'immensité de l'univers comme si la vision, encore fort partielle, que nous venons tout juste d'en acquérir, était d'ores et déjà acclimatée dans notre monde terrestre, replié sur la centralité. Je suis plutôt porté à croire personnellement que cette acclimatation est illusoire et même qu'elle n'aura vraiment jamais lieu. Pas plus que d'expédier avec d'infinies précautions étroitement terrestres quelques mouches, quelques souris, quelques singes, ou même quelques hommes en orbite autour de la planète ne constitue et ne constituera jamais une acclimatation réelle à l'espace interplanétaire.

En réalité, plus ou moins à notre insu, nous n'entrons culturellement et intellectuellement dans l'immensité "nouvelle" de l'univers qu'avec les "instinctives" précautions de survie imposées par notre condition d'origine, rivée à la centralité terrestre par la force considérable de ce que j'oserai appeler la "gravitation" culturelle de notre être et de tout ce qui nous entoure. Pas étonnant dès lors que, dans cette expérience limitée, nous nous retrouvions avec le sentiment d'un mal indéfinissable, qu'il faudrait sans doute appeler le "mal de l'immensité" (côté sombre de sa fascination), à l'analogie du "mal de l'espace" dont souffrent, paraît-il, à leurs moments de vérité, la plupart de nos courageux astronautes.

Je voudrais retenir ici le mot très significatif, et humainement vrai, de Blaise Pascal (1623-62): "Le silence éternel de ces espaces infinis m'effraie." On notera la triple perception conjointe, fondue en une seule, du silence, de l'éternité, et de l'infini. Silence et désarroi de l'interrogation proprement humaine, et silence encore plus épais de la réponse, ou balbutiement insaisissable de l'une et de l'autre.

En troisième lieu, il me semble utile d'attirer expressément l'attention sur ce que j'appellerai, pour faire court, la distinction du substantiel et du relationnel dans la conception fondamentale de l'être. Cette distinction, sans doute plus ardue à saisir que celles qui précèdent, me paraît en revanche toucher plus directement le coeur du problème qui nous occupe ici: celui des traces possibles d'une intentionnalité dans l'ensemble des phénomènes offerts à notre observation à la fois dans le monde et dans l'univers. En définitive, que cherchons-nous à discerner, en supposant, il va sans dire, que l'hypothèse elle-même ne nous égare pas dès le départ? – des traces d'ordre substantiel, ou des traces, probablement plus subtiles, qui seraient, par exemple, d'ordre relationnel. Posons donc ainsi la question, sans nous inquiéter outre mesure, pour le moment, de son caractère abstrait.

De ce point de vue, je remarquerais d'abord que la pensée occidentale, même celle qui se dit la plus scientifique, demeure à peu d'exceptions près la fidèle et docile héritière de la philosophie grecque. Or, le grand postulat de cette pensée, en ce qui regarde l'être, c'est que celui-ci appartient de soi à un ordre de réalité apparenté à ce qui, d'une manière générale, est lui-même assez *solide*, assez *consistant*, assez *résistant*, assez *durable* pour *soutenir* (c'est l'idée de substance, ou d'essence, saisissable par la pensée, donc dans une certaine mesure, contrôlable dans une définition arrêtée de la réalité qu'on envisage en chaque cas), – malgré tout, l'*insoutenable* précarité de l'*existence*, toujours menacée de *ne plus être*, toujours susceptible du même coup de nous frustrer de notre indispensable certitude de pouvoir jusqu'à un certain point demeurer maîtres de notre rapport au monde. Bref, dans la culture occidentale (héritière dominante de la pensée grecque), l'être se présente obscurément comme une réalité en sursis de non-existence: menacée en elle-même et par-là même menaçante pour moi. Justement, ce qui, en dépit de cette précarité fondamentale, permet à l'être d'apparaître dans le monde d'abord (les phénomènes), et de se présenter ensuite à l'observation d'un témoin de l'être tel que l'homme (on songe ici aux opérations éventuelles d'analyse et de définition), c'est sa *substance*, ou son *essence*: substance ou essence par laquelle je nomme l'être, et par laquelle aussi j'ai en quelque mesure prise sur lui (ordre de l'action, de la fabrication, et de la création).

On notera, du reste, que dans cette pensée, et dans notre culture plus généralement, le recours à la causalité, quelle qu'en soit la morphologie précise, suit le même modèle de substantialité première et fondamentale de l'être. La cause qui mérite son nom doit de quelque manière posséder la même solidité, la même consistance, la même résistance, la même durabilité que la substance, qui fait l'être lui-même, qui la pose au-delà du champ toujours menaçant et toujours menacé de la non-existence. On aperçoit d'ici, je pense, ce que cette vue de l'être peut donner, et donne généralement, en ce qui concerne les recherches sur l'origine et l'évolution de l'univers.

Dois-je ajouter que j'éprouve un profond respect pour ce modèle de pensée? Du point de vue de la longue et féconde histoire de la culture occidentale, on peut l'évaluer favorablement à ses résultats, y compris bien sûr ceux qui se présentent aujourd'hui à nous sous le nom de science, sans parler de la technologie qui lui est de plus en plus étroitement associée. Je respecte cette vision de la substantialité première de l'être, et du même coup, la vision de la sorte de substantialité corrélative des causes qui gouvernent l'être dans le cours changeant de l'existence. Nous ne pouvons oublier que c'est une telle vision de l'être qui, à partir de Socrate et de Platon surtout, a le plus activement engagé la pensée, puis la culture occidentales dans la voie de la recherche inlassable de la définition achevée, en toutes choses de la terre et du ciel: c'est-à-dire, dans la voie de la définition que je maîtrise, qui ne me trahira pas dans les moments critiques, dont je puis croire qu'elle enserre valablement le réel, et pour finir le met efficacement entre mes mains, suprême récompense de l'effort que j'ai déployé pour m'en assurer la conquête.

Je respecte ce modèle de pensée, mais j'ajouterai, si vous me permettez: comme je respecte, d'un point de vue anthropologique, l'image première de la centralité du "monde" où nous sommes, sans préjudice de la très importante correction que la science est maintenant en mesure d'apporter à cette image de base (il s'agit de rien de moins que d'une image première dans l'ordre des structures du rapport global de l'homme au monde).

Je m'explique brièvement. L'image première de la centralité s'impose d'elle-même dans le développement ontogénétique de chaque individu, comme elle s'est imposée d'abord, il va sans dire, dans l'évolution phylogénétique de notre espèce. Par ailleurs, l'image seconde de l'immensité de l'univers (dans laquelle je risque de perdre mon centre) me parvient, ou ne me parvient pas (c'est malgré tout aléatoire), par simple voie d'apprentissage, et la correction qu'elle m'apporte éventuellement ne peut rien faire d'autre, en réalité, que de coexister plus ou moins harmonieusement avec une image première, d'un autre ordre d'insertion dans la réalité humaine (ontogénèse, d'un côté, et connaissance, de l'autre), qu'elle ne peut prétendre ni recouvrir ni encore moins éliminer entièrement.

Cela dit, je me demande toutefois si aujourd'hui, compte tenu de l'état général de la culture, et notamment de la science (surtout physique), nous ne devrions pas entreprendre de corriger au moins partiellement le très ancien, très prestigieux et, dans son ordre, très efficace modèle de la substantialité première de l'être. Je crois, pour ma part, que nous sommes invités à entreprendre cette difficile opération de correction, essentiellement par suite de la "révolution atomique" (microstructures de base de ce qu'il est convenu d'appeler la "matière"), – un peu comme nous avons été forcés il y a déjà quelque temps de corriger l'image première de la centralité du monde (image contemporaine de l'homme lui-même, *homo sapiens*) en présence du dévoilement progressif de l'immensité toujours croissante (semble-t-il) des innombrables systèmes cosmiques dans lesquels prend place notre modeste système solaire, avec la planète encore plus modeste qui nous porte.

Corriger l'imposante et profonde image de la substantialité de l'être et de ses causes (culture occidentale), il me semble donc qu'il faut aujourd'hui l'envisager expressément, et non seulement dans les à-côtés de l'observation, de l'analyse, et de l'interprétation globales du réel. Mais alors, dans quelle direction? ou selon quelle morphologie en partie nouvelle des rapports fondamentaux de l'être et de la pensée?

Je dois par la force des circonstances me contenter ici de suggérer. Je ne démontrerai pas, et je ne plaiderai pas. Je laisserai plutôt l'idée, ou l'hypothèse elle-même exercer sa propre séduction, s'il y a lieu, sur vos esprits.

Au-delà de la substantialité apparente de l'être, je forme donc l'hypothèse que le vrai fond de l'être, sa vraie "nature," ou sa vraie structure, si j'ose dire, soit d'ordre *relationnel*. En d'autres termes, ce qu'il conviendrait de rechercher en premier et en dernier lieu dans l'observation, l'analyse et l'interprétation du réel, celui de l'univers comme celui du monde où nous sommes, je dirais que ce sont des *relations*, de morphologie variée, qui auraient leurs propres titres à l'existence, et des relations qui, en outre, posséderaient de tels titres de façon relativement stable.

J'inverserais donc ici les priorités *apparentes* de la pensée (surtout occidentale). Au lieu de chercher d'abord, et constamment, des objets substantiels (au sens déjà précisé), particules ou galaxies, objets à "voir," objets à "définir," il me semble que notre compréhension de l'ensemble de la réalité gagnerait en clarté (sinon en "évidence" première), en force et en cohérence, si nous acceptions de nous tourner résolument en premier lieu vers les "invisibles" *relations* qui tissent tout le fond (sans exception) de la réalité du monde et de l'univers, sans passer par le détour ("logique"?) de la substantialité (supposée première: hypothèse inverse) des objets dont nous nous appliquons ensuite à dévoiler les relations (plus ou moins étroitement codifiées en lois du monde et de l'univers). Bref, sans parodie, je dirais: au commencement était la relation: le reste a suivi, "substance" incluse.

Mais alors j'ajoute: si la relation était au commencement, j'imagine qu'elle doit, ou qu'elle devra également se retrouver à la fin. Ce qui signifie, pour le problème qui nous occupe: le tissu réel de l'origine et de l'évolution de l'univers se présente d'emblée, dans toutes les directions, comme un tissu relationnel. D'autre part, je ferais remarquer à ce point de notre analyse, que nous ne pouvons plus dès lors parler de l'*unité* de la relation dans le sens si longtemps entretenu, à travers à peu près toutes les discussions de la philosophie occidentale, autour de ce qu'on appelait depuis Parménide l'*unité* de l'être.

En fait, pour aller jusqu'au bout de l'hypothèse que j'essaie de présenter ici, il faudrait plutôt dire: la morphologie propre de l'*unité relationnelle*, c'est d'être *systémique*, ou de se présenter sous forme de système, quelles que soient d'ailleurs la simplicité ou la complexité du système lui-même selon les cas. On parvient ainsi au paradoxe suivant (qui n'est pas une contradiction): le propre de l'*unité* relationnelle, c'est d'être *multiple*, au lieu que l'unité de l'être (substantiel) s'accommode plutôt mal de la rencontre et de l'observation du multiple, et tend en fait à l'éliminer, au moins dans la pensée, pour ne retenir finalement que l'"unique."

Je précise d'ailleurs tout de suite, avant de passer à la considération expresse de la directionalité et de l'intentionnalité, que si nous nous posons ici la question de savoir s'il n'existerait pas des "traces d'intentionnalité" (evidence for *design*) repérables dans l'origine et l'évolution de l'univers, c'est il me semble en présence d'un monde et d'un univers d'emblée *relationnels* que nous devons nous placer en premier lieu, plutôt que devant un monde et un univers d'êtres substantiels, conçu d'après le modèle occidental.

Je viens de suggérer la possibilité qu'en son fond, sinon dans ses apparences immédiates, l'être de notre monde, et par voie d'analogie, l'être de cet immense univers où nous sommes si curieusement logés, soit de nature relationnelle plutôt que substantielle. Entre autres choses, cela pourrait vouloir dire concrètement, par exemple, que ce ne sont pas les "particules" (notez le terme) comme telles qui sont de première importance dans la constitution, et la considération de l'assise originelle de l'univers, mais la *liaison relationnelle* qui les retient les unes *en présence* des autres, selon des lois de réciprocité qu'il n'est pas impossible de mettre au jour.

Disant cela devant un auditoire comme le vôtre, je me retrouve ici un peu effrayé de mon audace; vous me pardonnerez au besoin la naïveté qui l'inspire, si tel est votre sentiment. Mais je me dis qu'après tout, notre conception courante de la substantialité première de l'être, fondée principalement sur le "solide" et le "visible," ne va pas elle non plus sans causer quelques maux de tête à ceux qui s'aventurent à en explorer les fondements. Naïveté pour naïveté peut-être, je me résigne donc d'avance au triste sort qui s'abattra bientôt sur moi. Mais avant de toucher cette fatale échéance, j'aimerais en dépit de tout préciser un peu mieux ma pensée sur ce point, afin de périr s'il le faut plus justement.

Faisons donc provisoirement l'hypothèse que l'être soit de nature relationnelle, et misons pour le moment sur l'invisible. Je fais donc le pari que s'il est à tout hasard possible de relever quelque trace d'intentionnalité dans notre monde et dans l'univers, cette trace, de toute façon, ne sera pas *imprimée* sur tout ce qui *apparaît* de l'existence. Il s'agira très certainement de quelque chose qui ne sera pas de l'ordre de l'apparence des phénomènes.

J'ajoute alors: si l'être est de nature relationnelle, le plus petit système qui le constitue ne peut posséder moins de deux termes, et ces termes, selon la loi fondamentale des systèmes, ne peuvent à leur tour être en état d'indifférence l'un par rapport à l'autre (loi de non-indifférence réciproque de tous les éléments de n'importe quel système). En ce sens, l'être est déjà à la fois un et multiple, obligatoirement, jusqu'en son modèle le plus simple. La relation, pour demeurer pensable, et je présume, pour accéder à son existence de relation, doit être envisagée au minimum entre deux termes.

Or, c'est ici, il me semble, que la chose commence à se corser, et à devenir intéressante. Car le plus petit système relationnel ne peut reposer sur deux termes *à tous égards* identiques l'un à l'autre. En ce sens, il n'y a pas de relation constitutive (je ne parle pas des relations simplement observées) qui serait portée de côté et d'autre par *le même*. Les vrais termes relationnels, dans la structure de base, sont donc non seulement multiples mais *divers*, au moins sous un aspect. Bref, ils sont, dirait-on, de "signe" différent, quel que soit le signe.

Si nous reportons maintenant ce modèle relationnel simplifié sur l'ensemble de ce qui existe, nous obtenons un monde où le divers et la différence, l'*autre*, peuvent se répandre et se diffuser à l'infini.

Je dis: se répandre et se diffuser. L'être relationnel dont nous avons fait l'hypothèse ne donne pas, en effet, un monde premièrement statique ("solide") qui se mettrait par ailleurs en mouvement, qui rattraperait en quelque sorte le mouvement en cours d'existence, mais un monde qui est d'*abord* énergie et mouvement: énergie et mouvement *porteurs* l'un et l'autre de l'*infinité des relations possibles*. Il s'agit alors d'un monde dès le départ, et à tout moment, contemporain, et de son origine, et de son évolution. C'est simplement notre façon de voir les choses qui nous conduit à séparer, et même à isoler en quelque sorte, dans le temps, l'origine et l'évolution de l'univers. La contemporanéité constante de l'origine et de l'évolution, suggérée par le caractère relationnel de l'être, me paraît offrir ici une vue, plus difficile sans doute, mais aussi plus "élégante" et plus juste, de la réalité mouvante qui nous entoure.

Mouvante, et pourtant relativement stable. J'insiste: relativement, mais non absolument stable. Les lois que nous avons longtemps jugées "rectilignes" et infrangibles, "commettent" elles-mêmes des infractions. Un monde de systèmes relationnels, quelles que soient leur simplicité ou leur complexité, est plus improbable qu'un monde construit sur le modèle substantiel, qui commence (pour asseoir son existence) par se consolider en nécessité et en certitude. Comme si, paradoxalement, il y avait plus de ressources de nouveauté dans le défini, l'arrêté et le certain, que dans l'océan illimité de l'improbable et du possible.

C'est ici qu'entre effectivement en jeu ce que j'ai appelé la directionalité des phénomènes observables dans le monde et dans l'univers. C'est, pourrait-on dire, la pièce maîtresse de l'ordre et de l'organisation du cosmos comme tel. La directionalité dont je parle n'implique d'ailleurs pas au premier stade l'idée de "but" ou de "fin," encore moins celle d'"intention" (*purpose, design*), qu'il faut se garder d'introduire trop tôt, ou trop facilement, dans l'observation, l'analyse et l'interprétation du monde qui nous entoure. But et fin sont des catégories d'*interprétation* de la direction ou de la directionalité: ce ne sont pas en premier lieu des catégories d'*observation* et d'analyse immédiates du réel (ou de la réalité du mouvement). Catégories d'interprétation (herméneutiques) très anthropomorphiques, d'ailleurs.

La directionalité ne dit en réalité que ce qu'elle dit: elle désigne un caractère général du mouvement comme tel, analogue à la temporalité et à la spatialité. Négativement, d'autre part, la directionalité nous avertit que le mouvement, dans toutes les tranches de réalité où il devient de quelque manière observable, n'est pas soumis à une prévalence, ou à une priorité, initiales du hasard. Elle nous invite à penser plutôt que l'ordre et l'organisation qui naissent du mouvement et dans le mouvement, ne sont pas rattrapés après coup, et en quelque sorte éventuellement *sauvés* de la totale indifférence du hasard. Elle nous souligne que la direction dans le mouvement reflète, en définitive, au plan observable des phénomènes, la non-indifférence fondamentale qui gouverne les rapports réciproques de tous et chacun des éléments relationnels d'un même système.

Pouvons-nous, parvenus à ce point, faire un pas de plus, et commencer à parler, non seulement de directionalité (ce qui engage déjà beaucoup), mais encore d'intentionnalité (ce qui engage infiniment plus à tous égards)? Il faut au moins avancer avec prudence dans un sens ou dans l'autre, qu'on envisage plutôt une réponse négative, ou qu'on envisage plutôt au contraire une réponse positive. En tout état de cause, il serait vain autant que présomptueux de tenter en cette matière quelque démonstration que ce soit. En avons-nous d'ailleurs un si grand et si absolu besoin? Chacun devrait savoir par expérience qu'il est d'autres voies que le savoir (ferme et contrôlé) pour parvenir à certaines réalités, parmi les plus hautes et les plus satisfaisantes qu'il nous soit donné d'approcher.

Quoi qu'il en soit, au surplus, je ne puis parler ici que pour moi-même. Nous entrons, avec l'intentionnalité, dans un domaine de compréhension du monde (je ne parle pas de la seule explication), où la part de l'expérience personnelle, intransmissible et intraduisible pour la plus grande et peut-être la meilleure part, joue par la force des choses un rôle de premier plan, fort difficile à évaluer d'ailleurs pour tous et chacun d'entre nous.

Timidement, et fort modestement, je dirais donc à peu près ceci en ce qui regarde les traces possibles d'intentionnalité que nous pourrions identifier, d'une part, dans notre connaissance et notre expérience globales du monde, et d'autre part, dans notre connaissance beaucoup plus abstraite, et plus récente, donc encore faiblement intégrée (culturellement), de l'univers dont nous faisons aussi partie.

Je le rappelle pour fins de clarté dans ma démarche. D'une part, nous trouvons, nous observons, la directionalité des phénomènes du monde et de l'univers. C'est essentiellement et en premier lieu un objet d'observation, qui pourra à un stade ultérieur de l'interrogation devenir un objet d'analyse, et à un stade encore plus avancé, un objet d'interprétation: la partie la plus risquée de tout le processus.

D'autre part, nous observons également dans le monde la présence, insolite à certains égards, disons même improbable, du point de vue de l'évolution globale de la vie planétaire, présence pourtant certaine, d'un être qui se présente d'emblée, non seulement comme un *utilisateur de la directionalité générale* des phénomènes, mais comme un *créateur d'intentionnalité*: d'une intentionnalité dirigée elle-même vers une communication avec le monde (curiosité, interrogation, etc.), ou vers une intervention dans le monde (action, opération, création, etc.), ou vers les deux à la fois. Ce créateur d'intentionnalité, c'est l'homme, vous et moi, depuis notre vieux frère, l'homme de Néanderthal.

Avec notre espèce, l'intentionnalité s'est pour ainsi dire installée dans la vie planétaire, notons-le avec soin, de façon ininterrompue. Ce dernier trait, en effet, est à lui seul autrement plus significatif, du point de vue de la structure propre du monde, et peut-être aussi de l'univers, que s'il ne s'était agi que de l'apparition d'un émetteur *occasionnel* de "flash" d'intentionnalité (qu'est-ce que cela pourrait bien être, de toute façon?), sorte de météore fortuit dans le ciel nocturne de la présence "directionnelle" au monde, situation obligée de tout le reste de ce qui nous entoure, situation déjà extraordinaire et admirable d'ailleurs.

Créateur d'intentionnalité, l'homme devient du même coup créateur de ses propres projets (ordre de la promesse), au-delà et sur la base, de tous les systèmes relationnels *programmés*, dont il repère la présence et l'action en lui-même et dans le monde. Projets et programmes: deux géométries inversées, pourrait-on dire, d'une même présence fondamentale au monde, un peu à la manière de la géométrie aérienne (ascendante) du tronc et des branches de l'arbre, inversée par rapport à la géométrie (descendante) souterraine, invisible et beaucoup plus improbable des racines du même arbre cherchant à se frayer à tâtons un chemin nourricier dans le sol: géométries inversées, semblables et pourtant profondément différentes de la même et unique présence vivante de l'arbre au monde (milieu).

Projets formés jour après jour dans l'intentionnalité restreinte d'une présence individuelle ou collective au monde. Programmes établis depuis des millénaires dans la directionalité générale de la présence de notre espèce au monde. Projets et programmes: liaison puissante d'une complexité inouïe, dont nous nous rendons à peine compte, et qui est certainement unique sur notre planète, même s'il est plausible de supposer que des liaisons analogues pourraient se retrouver ailleurs dans le vaste univers. Projets et programmes: liaison fondée en dernier lieu, sur l'acquisition, autour de ce qu'il est convenu d'appeler le seuil de l'hominisation, d'un dispositif remarquablement efficace de représentation et de prospection du monde: le *symbole*. Après l'apparition de la vie elle-même, l'accès de notre espèce à la représentation et à la prospection symboliques du monde me paraît être l'événement, ou mieux l'*avènement*, le plus significatif à lui seul, de toute l'histoire de la planète.

Or, le propre de la représentation symbolique, par comparaison avec les deux régimes antérieurs, et plus archaïques, du signal-signe et de l'image-guide, c'est de mettre celui qui en dispose en mesure de *traiter la distance comme distance* dans son rapport au monde, cette distance devenant dès lors (dans l'ordre des possibles) virtuellement illimitée dans l'espace comme dans le temps.

Je ne puis m'étendre ici sur ce point pourtant capital, mais je pourrais dire brièvement, d'une part, que le régime de représentation que j'appelle image-guide (le plus archaïque des trois, dans le monde vivant) fonctionne dans le *contiguïté*, et ainsi dans une situation en tendance vers le degré zéro de la distance dans l'espace aussi bien que dans le temps; et d'autre part, que le régime plus évolué du signal-signe fonctionne dans la *proximité*, et alors dans une proximité dont le champ d'expansion maximale se limite toujours, et nécessairement, aux possibilités physico-chimiques concrètes des signaux et des signes utilisés en chaque situation donnée. En regard de ces deux régimes plus anciens, le trait distinctif, et véritablement extraordinaire du régime du symbole, c'est d'avoir permis à l'homme, et à lui seul, de *lever les frontières* étroites de la contiguïté, puis de la proximité, *pour passer au traitement régulier de la distance comme telle*, sans limites assignables ni dans l'espace ni dans le temps. Il est bien clair qu'en l'absence d'un tel dispositif de traitement de la distance comme distance dans la représentation et la prospection du monde, nous ne serions pas ici pour nous interroger sur l'origine et l'évolution de l'univers. Les tulipes et les hirondelles nous sont revenues avec le printemps, mais elles ne sont pas apparemment très intéressées par notre colloque. Il est vrai que nous avons omis de leur en parler! Question de traitement de la distance en communication, analogue à celle que j'évoquais à l'instant à propos de la représentation et de la prospection.

Pour résumer, et pour conclure, je reprends donc ici la formulation de la donnée fondamentale. L'homme ne se présente pas dans le monde comme un simple utilisateur, plus habile, de la directionalité générale du monde et de l'univers. Il se présente comme le créateur d'une donnée nouvelle qui comprend, d'une part, l'intentionnalité continue de son projet culturel global à travers l'histoire, et d'autre part, en ce qui concerne l'individu, l'intentionnalité quotidienne et inlassable d'un projet personnel, non-réitérable et unique, établi sur la base commune de la directionalité de tous les programmes qui entrent dans la composition de son système global.

Mais alors, je pose la question sur laquelle je voudrais conclure, et que sans doute vous attendez depuis un bon moment avec impatience:

quel peut bien être le sens de la présence dans le monde (le nôtre), présence tardive en apparence, d'un *créateur* d'intentionnalité individuelle et collective tel que l'homme?

Pourrait-on parler, par exemple, en partant de l'homme et de son rapport étroit avec la directionalité générale du monde et de l'univers, de quelque chose comme une intentionnalité première, plus ancienne, plus puissante aussi, dont nous pourrions peut-être apercevoir une trace dans l'évolution même de la vie planétaire? A moins de supposer plus simplement (en apparence) que l'avènement, qui nous paraît significatif, de l'intentionnalité humaine, puisse être une forme nouvelle, seulement un peu plus "évoluée," de la directionalité générale ou restreinte du monde et de l'univers, assise commune (cosmique) de tous les phénomènes. Auquel cas, l'interrogation s'arrête, du moins en apparence, ou va se diluer dans le hasard, comme une vaine émanation passagère de notre cerveau.

Mais, il y a un moment, nous avons cru devoir noter au passage, que la directionalité générale ou restreinte des phénomènes observables dans le monde et dans l'univers est incompatible avec l'idée d'une prévalence anticipée du hasard.

Peut-on alors aller encore plus loin, et parler d'une intentionnalité absolument première? Cette intentionnalité première serait, dans ce cas, je suppose, celle d'un être, ou d'un sujet, d'ordre relationnel autre que le monde, instaurateur véritable de la directionalité de base que nous observons dans le monde et dans l'univers; instaurateur *également, au-delà et au-dessus de cette directionalité de base*, d'une intentionnalité d'abord *diffuse* dans l'évolution même la vie planétaire; puis d'une intentionnalité *différenciée* (acquisition phylogénétique du symbole comme dispositif essentiel du rapport de l'homme au monde), reconnaissable donc à ce titre, et repérable en fait dans l'homme créateur d'intentionnalité: celle du projet culturel de l'espèce et celle du projet personnel de chacun.

Dans la façon dont j'essaie de voir présentement les choses, ce qui me paraît ici le plus significatif, sinon décisif, ce n'est pas l'intentionnalité comme telle, ou observable comme telle dans le présent, c'est plutôt, à un certain stade de l'évolution générale de la vie planétaire, l'*avènement d'un incontestable créateur d'intentionnalité*, et cela, dans un monde entièrement régi, du moins au plan des phénomènes immédiatement observables, par une directionalité généralisée, commune autant que nous sachions à l'ensemble de l'univers.

Je ne vois pas, en effet, comment l'avènement d'un créateur d'intentionnalité, si limité qu'il ait pu être, pourrait aujourd'hui s'expliquer, sans faire l'hypothèse (c'est l'inférence à mon avis la plus raisonnable) de la présence "ailleurs" que dans notre monde, et probablement aussi, "ailleurs" que dans l'univers (je dis "ailleurs" pour dire "autre," et faute d'un meilleur terme), d'une *intentionnalité qui*, elle, *n'aurait pas besoin d'explication* ni quant à son origine, ni quant à son existence, ni quant à sa relation avec le monde et avec l'univers où nous sommes. Bref, le recul de l'explication, pour demeurer utile et "raisonnable," doit s'arrêter quelque part.

Reste évidemment à rechercher, par une autre voie, quelle *représentation* nous pourrions éventuellement nous donner de la présence ailleurs que dans le monde et dans l'univers d'une intentionnalité qui ne serait pas soumise à nos conditions d'existence. Je dois laisser à chacun de vous le soin d'en juger par lui-même.

JAMES E. LOVELOCK

The Ecopoiesis of Daisy World

Truth is said to be stranger than fiction, and this may be why fiction is often more credible than fact. For example, anyone interested in the sociology of Victorian England could read the first great sociologist, Marx, but is more likely to have read Dickens. These thoughts were brought home to me recently following the publication of a book of fiction entitled *The Greening of Mars* by Allaby and Lovelock (1984). I was surprised by the seriousness of its reception. It was the story of an entrepreneur who made a profit by making the climate of Mars favourable for life. He used cast-off ICBMS and other surplus military hardware to carry to Mars the material needed to transform the environment and make it fit for life. Once it became warm the planet was seeded and allowed to develop its own ecosystem. It eventually became a place where colonists could make a home that was independent of support from the Earth.

This fiction was intended as entertainment, but it appeared by chance at a moment when many were giving serious thought to colonizing nearby planetary bodies. Among them was Robert Haynes, who visited me in 1984 to discuss what would be needed to establish an ecosystem on Mars. He brought with him his splendid newly minted word "ecopoiesis." The term refers to the fabrication of an ecosystem or biosphere on a lifeless planet, thereby establishing a new arena in which biological evolution ultimately can proceed independently of that on Earth. The making of a home for life. It seemed particularly well suited to describe our intentions towards Mars; more generally it is applicable to the study of ecosystems on a regional as well as a planetary scale.

It is easy enough in the science fiction of glossy magazines or in enamelled documentaries to build self-sustaining colonies in space or on the Moon or Mars. The daunting problem of a permanent life-support system can be naively dismissed as solved by an automatic ecological balance between humans as consumers and a few horticultural plants. It is exceedingly unlikely that this simple notion would work, and for practical space colonies we are left with an engineering approach, namely the life-support systems of space vehicles and nuclear-powered submarines. These have been developed to near perfection and work well in small-scale systems and where cost is not a limiting factor. A real self-sustaining human colony on Mars could not afford such a luxury and would have to live off the planet. It would be a great help to the colonists if Mars possessed a self-sustaining ecosystem, as does the Earth. Then the air conditioning, the climate regulation, and the supply of food would be an automatic background that required no thought or effort for its enjoyment.

The steps needed to make Mars a fit place for life to start are a reasonable subject for speculation and modelling; this can be done in spite of some gaps in our knowledge about the Martian surface. In particular we need to know how much water and carbon dioxide are readily available from the permafrost of the Martian regolith.

Much more important than the start of life on Mars is the acquisition by that life of the control of its planet. This is what ecopoiesis is really about and it is the area of our greatest ignorance. We have not even started to model the intricate interaction between the biota and its environment on Earth; to attempt it for the unknown conditions of Mars would be naively optimistic. What we can do is to follow the normal practice of science and reduce the problem to a reasonable size by abstracting what seem to be the essentials. Therefore I have chosen for discussion the ecopoiesis of a simple world where the environment is described by a single variable temperature and the biota limited to a single species, daisies. Daisy World is not only a fine place for experiments in ecopoiesis but is also one where other matters of interest relating to the themes of this volume can be discussed, such as the teleological argument and the colligative properties of life.

An intuition that life on a planetary scale is more than just a catalogue of species, or of the genes that they express, is ubiquitous. It has persisted from primitive times and it survived nineteenth-century reductionism. It drew occasional scientific sustenance from those pioneers like Boussingault and Dumas who recognized the dynamics of the cycling of the elements. In more recent times it led Vernadsky (1945) and Hutchinson (1954) to lay the foundations of biogeochemistry. It also led the first courageous ecologists like Odum (1968) to recognize the possibility that ecosystems had an identity and were to an extent self-regulating.

A planet-sized look at life was forced by the attempts in the 1960s to design experiments to detect life on Mars (Hitchcock and Lovelock 1966). This telescopic vision showed Mars to be lifeless, a fact confirmed ten years later by the Viking lander. More important the same external view of the Earth substantiated the intuitions of astronauts that earth was in an important sense alive. It led me to propose that the earth was a self-regulating system with the capacity to keep itself in homeostasis so that the climate and chemical composition were regulated by and for the comfort of the biota. It has been the subject of an enduring collaboration (Margulis and Lovelock 1975).

A detailed account of the large body of evidence we have gathered in response to tests of this, the gaia hypothesis, cannot be given here (see Lovelock 1979). It is testable and so far predictions based on it have been verified. Instead, let us go back to that simple planet Daisy World. It came into being in response to thoughtful and helpful criticisms by Ford Doolittle (1981) of the gaia hypothesis. He rejected the notion of planetary self-regulation since it would need foresight or planning by the biota, which was impossible. He also held that there was no way for natural selection to lead to the evolution of a global altruism. Other critics of gaia have condemned it as teleological, but their criticism is dogmatic. Many scientists are confused about the distinction between the false circular argument of classical teleology and the right, proper, and essential recursive logic of living things and of self-regulating mechanisms. It is good to have a sense of wonder about first causes and purpose in the universe, but usually we recognize these speculations to be untestable in a Popperian sense. It is inexcusable dogmatically to condemn as teleological the circular logic of a thermostat, an automatic pilot, or of gaia. The status of cybernetics as a science has suffered grievously from the prejudice that arises from this ignorance.

The Ecology of Daisy World

To answer Ford Doolittle's criticisms I developed a model planet on which the simple competitive growth of different coloured daisies was enough for a powerful and long sustained regulation of the planetary climate at a level comfortable for daisies. The model was tested by subjecting it to a wide-range heat input as would take place during the

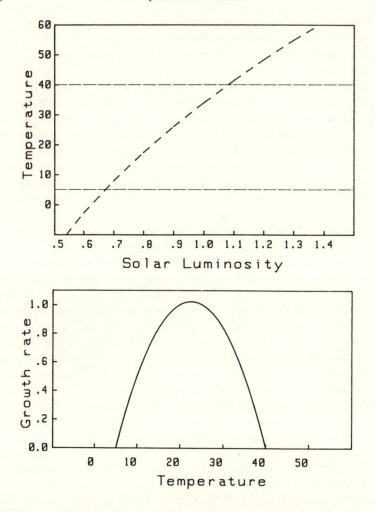

Figure 1 The calculated increase of mean planetary surface temperature (Curve A) during the evolution of a main sequence star. The stellar luminosity varies from 0.6 to 1.4 times the luminosity of the Sun and the temperature over the range -10 to 60 degrees Celsius. Also illustrated is the growth rate of vegetation (Curve B), taken to vary from zero to unity in a parabolic manner between 5 and 40 degrees Celsius.

normal evolution of a main sequence star. A more rigorous account of the mathematical basis of the model is to be found in Watson and Lovelock (1982).

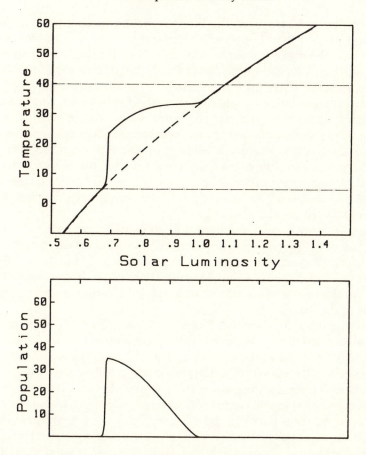

Figure 2 The change of mean surface temperature and plant population when the planet supports the growth of dark-coloured vegetation (albedo 0.2) and when the bare rock is more lightly coloured (albedo 0.4).

Imagine a planet similar to but with less ocean than the Earth; it travels at the Earth's orbital distance around a star of the same mass and composition as the Sun. The planet spins like the Earth and has a simple atmosphere with few clouds and no significant quantities of greenhouse gases. Its mean surface temperature is determined simply by the balance between the radiation received from its sun and that re-radiated to space. From a knowledge of the mean surface albedo it is easy to calculate the surface temperature (T):

$$T = (\text{Solar flux} \times (1 - \text{albedo})/ 4 \times \text{sigma})^{.25} \quad (1)$$

Sigma is the constant from the Stephan-Boltzman Law relating the emission of heat by a body to its temperature. Curve *A* of Figure 1 shows how the mean temperature of the planet would change during the evolution of its star on the assumption that the history of the star was the same as that of the Sun. A rare near certainty of astrophysics is that stars increase their output of heat as they grow older. Curve *B* of the figure shows how the rate of growth of vegetation would vary during this evolution of the planetary temperature. Most plants do not grow when the temperature is below 5 or above 40 degrees Celsius; they do best at about 22.5. In Figure 1 it is assumed that the growth of vegetation and the evolution of the planetary temperature are independent of one another. This has long been the conventional wisdom for the evolution of the Earth's climate.

Figure 2 shows how different the climate history is when the evolution of the planetary temperature is coupled with the growth of vegetation. This could happen if the colour, the albedo, of the vegetation was different from that of the bare rocks. In the figure the albedo of the vegetation is taken as 0.2, typical of forests on Earth, and that of the bare rocks 0.4.

The explanation of the difference is that when the planetary temperature reaches 5 degrees Celsius, some of the plants begin to grow. Since they are dark in colour and absorb heat, the region where they grow will be warmer than the surrounding still barren rocks. The warmth will accelerate the growth and the spread of the plants; in turn the planetary albedo will begin to fall as larger and larger areas become plant covered. In a powerful positive feedback the temperature will rapidly rise until the further advance of plant cover is deterred by the plants themselves becoming hotter than the best temperature for growth. The planetary temperature will stabilize at a level somewhat above the optimum. As the heat from the star increases the area covered by the plants will steadily decrease but the climate will keep constant until the plants have died.

In this model and in those that follow the growth and spread of the plants was taken to be described by the equation developed by Carter and Prince (1981) and verified by observations of daisies at a site a few miles from where I am now writing.

$$d(\text{daisies})/dt = \text{Beta} \times (\text{Bare earth} - \text{Gamma}) \quad (2)$$

Beta and Gamma are respectively the growth and death rates of the daisies. The models illustrate the powerful effect on planetary temperature of a combination of the physical radiation balance and

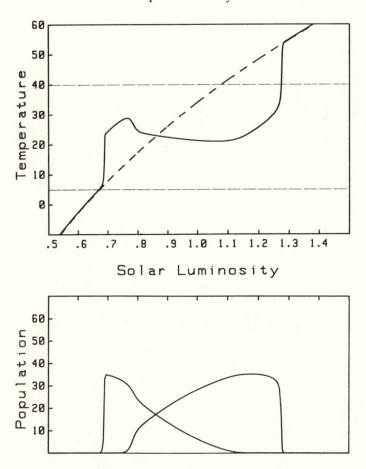

Figure 3 The regulation of planetary temperature when it supports the growth of two different daisy species. One dark (albedo 0.2) and one light (albedo 0.6). The variation of the population of these two species is also shown.

biological growth. No foresight or planning is needed by the plants, merely a difference in colour from that of the bare rocks.

Figure 3 illustrates a similar evolution of temperature and plant growth, but this time with two species of plants present, one darker and the other lighter than the bare surface. Here the thermostasis is much more effective and extends over a wide range of heat input from the star; in addition the total biomass remains constant over the range of thermostasis instead of declining as with a single colour in Figure 2.

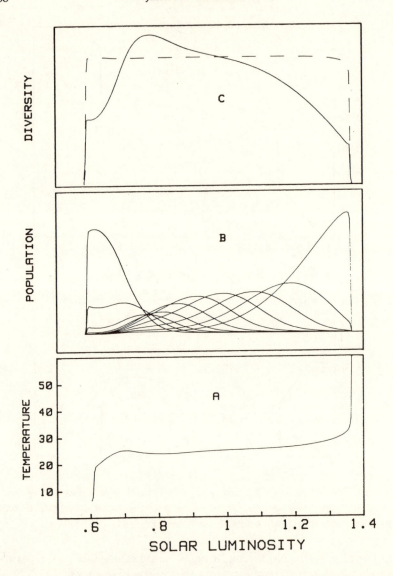

Figure 4 Thermostasis with ten differently coloured daisies growing in competition during the temperature evolution of the planetary system. Curve *A* is
the temperature history, Curve *B* the population distribution of the individual
species, Curve *C* the total biomass (dashed line) and the ecological parameter,
diversity (solid line).

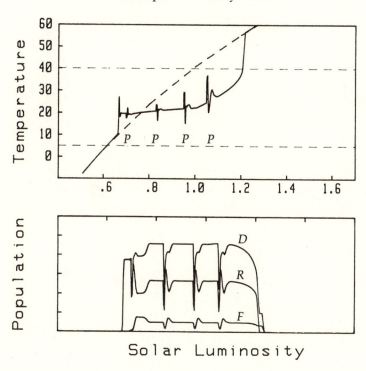

Figure 5 Thermostasis on Daisy World when there is a single species of daisy (*D*) able to change its colour adaptively and when there are rabbits (*R*) grazing them and foxes (*F*) culling the rabbits. At intervals, marked by (*P*), the model is perturbed by the sudden death of 30 per cent of the daisies. The variation of the population of the three species is illustrated in the lower part of the figure.

Figure 4a illustrates the temperature evolution of our imaginary daisy world when there are ten differently coloured species in competition for the occupation of the planetary surface. The variations of the populations of the different daisy species is in Figure 4b, and the total population (biomass) is shown in 4c together with the theoretical ecological parameter diversity.

$$\text{Diversity} = \text{Population } (i) \times \log (\text{Population } (i)) \quad (3)$$

Figure 5 is an attempt to bring daisy world closer to Earth by including on it rabbits to graze the daisies and foxes to eat the rabbits.

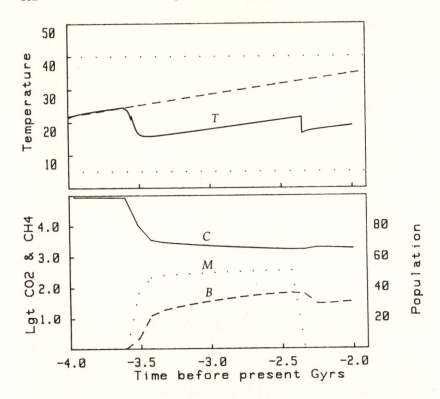

Figure 6 Thermostasis where the growth of prokaryotic photosynthesizers is a sink for atmospheric carbon dioxide and where the digestion of dead photosynthesizers by methanogens is a source of methane and carbon dioxide. Curves *T*, *C*, *M*, and *B* are respectively the variation of the temperature, carbon dioxide, methane, and biomass during the evolution of the model. It could serve as a speculation about the history of the archean period going from the start of life to the appearance of oxygen.

Again thermostasis is excellent, although this model also includes perturbations. At intervals marked by a *P* on the figure the daisy population was suddenly reduced by 30 per cent to simulate some natural disaster such as a plague or fire. It is interesting that the effect of these perturbations, although always resisted, steadily increased in severity as the system aged and thermostasis became more difficult.

Figure 6 shows temperature regulation achieved by an ecosystem that used photosynthesis to cycle carbon through the atmosphere as carbon dioxide and methane. Both of these bases are products of the biota, and their atmospheric abundance powerfully affects the radiation balance and hence the mean temperature of the planet. The figure serves to illustrate that the "Daisy World" model is not limited to flowers but can also be used with prokaryotes. It may also provide insight to the understanding of the geophysiology of the archean period on Earth and to the ecopoiesis of Mars.

Daisy World is fertile and it can be the site of other imaginary ecologies. It is a place where the unplanned adaptation and the unconscious competition of the species leads to planetary homeostasis.

The models share in common four essentials: (1) a biota whose species tend to grow rapidly to occupy any niche that opens for them; (2) Darwinian natural selection, in which the species that leaves the most progeny inherits the niche; (3) a biota whose growth is limited by physical and chemical constraints (it can be too hot or too cold, too acid or alkaline, or there can be too much or too little of a nutrient and so on); and (4) a physical and chemical environment affected by the presence of the biota.

Daisy worlds represent the cybernetic system illustrated in Figure 7. It is similar in form to those that describe a physiological process like the regulation of temperature in an animal. On Daisy World growth acts as an amplifier and natural selection as the sensor of the physical state. The alteration of the environment by the presence of the biota and the effect of the environment upon the biota serves to bind them tightly in a web of feedback. The evolution of life and the evolution of the physical environment are a single indivisible process.

The physiologist Walter Cannon proposed the term homeostasis for the special wisdom of the body that keeps it in a steady state. The automatic regulation of planetary properties seems similar to homeostasis in animals; it seems appropriate to apply the term geophysiology to describe the investigation of planetary homeostasis. It may seem radical to view the Earth as a living planet, yet the objection of geneticists, who see action at such distance to be an impossible extension of the phenotype, is understandable but wrong. The life and the earth sciences are still immersed in paradigms that fail to recognize the close coupling between life and its physical environment. It is like the analogous tendency of an innocent quantum molecular chemist to disregard gravitation as a significant force. It takes a planet full of matter for gravitation to be noticeable. It takes a planet full of life for the existence of a geophysiological system. In the context of this sympo-

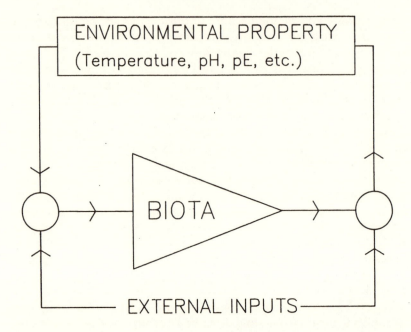

Figure 7 A diagram drawn from control theory to illustrate as a single system the active feedback between the biota and its environment. The biota is represented as an amplifier connected to a sensor that recognizes any departure from the operating point of the system. Physical and chemical variables, such as temperature or oxygen concentration, are summed and compared with the operating point of the system. If there is a difference the biota responds by active feedback to oppose it and so keep the system in homeostasis. The system also has the capacity to evolve by moving the operating point to a new steady value. This form of system evolution is called homeorhesis.

sium we might wonder if the universe, as the largest collective, also has unexpected bulk properties that are not yet visible.

Daisy worlds have vistas interesting to other eyes. Theoretical ecologists may have noticed that something unusual is illustrated in Figures 4 and 5, where as many as ten species are modelled in comfortable coexistence. Previously to model the competition of more than three species was impossible because of mathematical instability. There has been no shortage of effort during the forty years following the pioneering of Lotka and Volterra; what has been lacking was the stabilizing influence of feedback from the physical environment. An engineer, if asked, might have seen that theoretical ecology was running its models open loop, and warned that this is a notoriously unstable procedure.

Also, some of the disagreements about the interpretation of Darwin's theory can be resolved if evolution is taken to be a single process that includes life and its geological environment as a single tightly coupled system. Geophysiology predicts that in these circumstances evolution would proceed, with long periods of stasis interrupted by punctuations. These might come from some sudden external force or arise from an internal contradiction. In either event the system would quickly move from one stable state to another. This may be why Gould and Eldredge (1977) are right to postulate that evolution is punctuated rather than gradual.

What has all this to do with ecopoiesis, you ask? I think that it has a great deal to do with it. It suggests that sparse life existing in a few oases could never survive to control a planet. The ineluctable forces of the physical and chemical environment would overwhelm it. It offers choice for those who would terraform a planet: there is planetary engineering, the ceaseless task of keeping a planet air-conditioned, supplied with water and with nutrients; or there is ecopoiesis, the making of a planetary home for life and then leaving it to evolve, so that the climate and the chemical composition are regulated automatically at what is desired, and for free. Ecopoiesis, if successful, is a continuing process, like marriage. As members of the present system on Earth we have these same choices for our relationship with our planet. We can be the conscripted crew of that unstable contraption the spaceship earth, forever enslaved as planetary maintenance engineers; or we can continue to be partners of the gaian enterprise that has served so well for 3.5 billion years.

WORKS CITED

Allaby, M., and J.E. Lovelock. 1984. *The Greening of Mars*. New York: St Martin's Press.

Carter, R.N., and S.D. Prince. 1981. "Epidemic Models Used to Explain Biogeographical Distribution Limits." *Nature* 213:644–5.

Doolittle, W.F. 1981. "Is Nature Really Motherly?" *CoEvolution Quarterly* 29:58–63.

Gould, S.J., and N. Eldredge. 1977. "Punctuated Equilibria: The Tempo and Mode of Evolution Reconsidered." *Paleobiology* 3:115–51.

Hitchcock, D.R., and J.E. Lovelock. 1966. "Life Detection by Atmospheric Analysis." *Icarus* 7:149–59.

Hutchinson, G.E. 1954. "Biochemistry of the Terrestrial Atmosphere." In *The Solar System*. Ed. G.P. Kuiper. Chicago: University of Chicago Press.

Lovelock, J.E. 1979. *Gaia: A New Look at Life on Earth*. Oxford, New York: Oxford University Press.

Margulis, L., and J.E. Lovelock. 1976. "The Biota as an Ancient and Modern Modulator of the Earth's Atmosphere." *Pageoph* 116:239–43.

Odum, E.P. 1968. "Energy Flow in Ecosystems: A Historical Review." *American Zoologist* 8:11–18.

Vernadsky, V. 1945. "The Biosphere and the Noosphere." *American Scientist* 33:1–12.

Watson, A.J., and J.E. Lovelock. 1982. "Biological Homeostasis of the Global Environment: The Parable of Daisy World." *Tellus* 35B:284–9.

JÁN VEIZER

The Earth and Its Life: Geologic Record of Interactions and Controls

Introduction

The search for design in our surrounding world is strongly influenced by personal views of the relative importance of deterministic vs probabilistic concepts of evolution and history. The deterministic attitude naturally leads to formulation of an ultimate cause, and thus of a directional design. The probabilistic approach, although not excluding design in terms of processes rather than products, does not require directionality. I would like to argue, from a geological perspective, that these two concepts are not mutually exclusive, but complementary. The evolution of the Earth can be viewed as a propagation of populations through continuous "birth/death" cycles. This concept is true whether the "dead" inorganic or the "living" organic matter is considered. The populations, however, are mutually interdependent and of variable size. On average, the larger the population the longer its life-span. The "birth/death" cycles of large populations establish the limits, and the very basis of existence, for the operation of smaller populations. For subordinate populations, departures from such controlled steady-states are possible, but only on time scales shorter than the life-spans of the dominant populations. Thus an event such as habitat destruction may be of deterministic significance for a given living community, while at the same time being of only repetitive (probabilistic) significance for the controlling geologic cycle (for example, creation or destruction of mountain ranges). This interaction of "birth/death" cycles, or population dynamics, will be a recurrent theme of the subsequent discussion.

Conceptual Approach

An ideal natural population, characterized by a continuous "birth/ death" cycle, is usually typified by an internal age distribution pattern similar to that in Figure 1. The proportion of progressively older constituent units, such as human individuals, decreases exponentially. This exponential (power law) decrease results from mortality being usually proportional to, or a first-order function of, the size of the given age group. A cumulative curve of such an age histogram defines all necessary attributes of a given population. These are its *size A* (here 100 per cent), *half-life* τ_{50}, *mean age* τ_{mean}, and *oblion age* (or maximal life-span) τ_{max} (Fig. 1). The above τ_{50}, τ_{mean}, and τ_{max}, and thus the slope of the cumulative curve, are an inverse function of the mortality rate; the faster the rate the steeper the slope, because older individuals have a lesser chance of survival. For a steady-state extant population, natality per unit time equals combined mortality for all age groups during the same time interval. Consequently, the cumulative slope remains the same but propagates into the future (Fig. 1 bottom). This natality (= mortality) rate is designated as the *average recycling proportionality parameter b*. The above terms constitute the minimal terminology needed for geologic application of the concept, usuall designated as *population dynamics*. Further details are available in [11, 14, 21, 28, 30, 31, 36, 45, and 48]. The subsequent treatment of the solid earth, and of its hydrosphere, atmosphere, and biosphere is based on this unifying dynamic concept and on the sets, or hierarchies, of populations.

Hierarchy of Geological Populations

According to the theory of global plate tectonics [13, 24, 29, 34, and 44] the upper layer of the solid Earth, the crust, is subdivided into several major global tectonic realms (Fig. 2). These are:
- (a) basaltic oceanic crust and its overlying deep sea sediments;
- (b) active margin basins separated from (a) by tectonic barriers (e.g., Sea of Japan, Caribbean, Mediterranean);
- (c) passive margin basins (e.g., the Atlantic continental shelf and slope);
- (d) immature orogenic (mountain) belts, such as the Andes or the Rocky Mountains;
- (e) mature, or worn-down, orogenic belts (the Appalachians and the Hercynian mountain ranges of Europe represent imperfect, not yet entirely worn-down, examples);

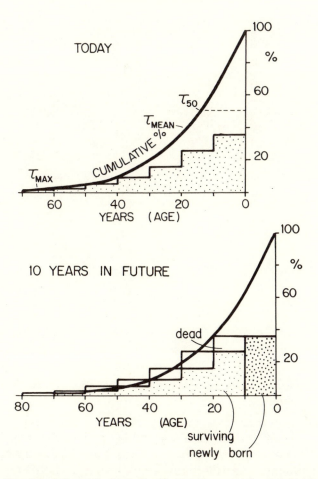

Figure 1 Simplified theoretical age distribution for a steady-state living system. τ_{50}, τ_{mean}, and τ_{max} as in the text. In this case, the natality/mortality rate is 35% of the total population for a 10-year interval ($b = 35 \times 10^{-3}$ year^{-1}). τ_{max} is arbitrarily defined as the fifth percentile. In practice, τ_{max} marks the point where the resolution of the data base becomes indistinguishable from the background. Reproduced from Veizer and Jansen [51].

(f) platforms, that is, undisturbed flat lying sediments deposited on the stabilized crystalline basement of the continental crust (e.g., the North American midwest or the Russian platform); and

(g) cratons, here understood as the crystalline and metamorphic basement of continents (e.g., the Canadian Shield).

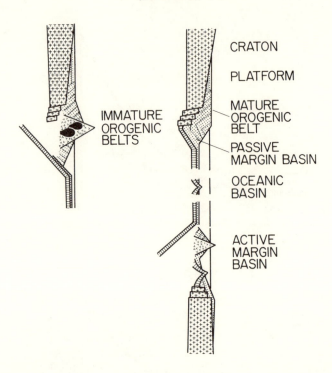

Figure 2 Schematic presentation of major tectonic realms in the context of the global plate tectonics.

These tectonic realms were selected because of the availability of quantitative data concerning the mass-age or area-age distributions for their constituent rocks. Alternative tectonic classifications are possible, but would not be especially dissimilar from the one above. In general, the tectonic realms (a) and (b) are mostly developed on oceanic crust, whereas (c) to (g) are associated with the crust of continental (granitic) type.

The mass-age or area-age distributions of rocks in these tectonic realms have been discussed in detail in Veizer and Jansen [53], and here only the distributions for continental and oceanic crusts are reproduced for illustration (Fig. 3). The cumulative curves for age distributions of rocks mimic the age distributions in living systems and both conform with the concept of population dynamics. The disparities between the living human and the "dead" inorganic populations are mostly in their sizes (10^{14} vs 10^{24}–10^{26} grams) and life spans (10^1 vs 10^7–10^9 years), but not in the age distribution patterns.

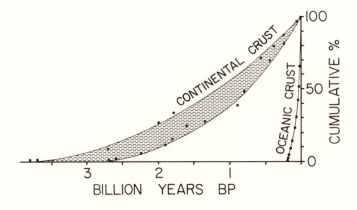

Figure 3 Area-age distributions of oceanic and continental crusts. For age distribution patterns of other tectonic realms see Veizer and Jansen [53].

The summary of the observed theoretical half-lives (τ_{50}) and recycling (natality/mortality) rates is given in Figure 4. This figure demonstrates good agreement between the theory and observations, with half-lives diminishing as the rate of recycling increases. The calculated half-lives for the particular tectonic realms are the following: basins of active margins ≈ 27 Ma (million years), oceanic sediments ≈ 51 Ma, oceanic crust ≈ 59 Ma, basins of passive margins ≈ 75 Ma, immature orogenic belts ≈ 78 Ma, mature orogenic belts ≈ 355 Ma, platforms ≈ 361 Ma. The τ_{50}s for continental basement depend on the resistivity of the particular isotopic dating system to later resetting, and range from ≈ 673 Ma for K/Ar, through ≈ 987 Ma for Rb/Sr and U-Th/Pb, to <1728 Ma for Sm/Nd isotopic pairs. The maximal life-spans τ_{max} for these tectonic realms are usually 3.0-3.5 times their respective τ_{50}.

The calculated data appear to separate the global tectonic realms into two domains, the continental and the oceanic one. The former comprises tectonic realms (e) to (g) and the latter (a) to (d). The oceanic domain, with τ_{50} in 10^7 years range, represents a relatively fast cycling system, while the more massive continental domain operates more slowly on time scales of 10^8–10^9 years. These two domains, or two interlocking wheels of different sizes and revolving speeds, interact in the process of mountain formation and destruction (orogenesis). Yet, orogenesis itself is a response to the operation of the oceanic cycle, as attested by recycling parameters for immature orogenic belts (Fig. 4). Thus, the generation of oceanic crust from the mantle, its spreading, and the peripheral mountain building processes are all interrelated

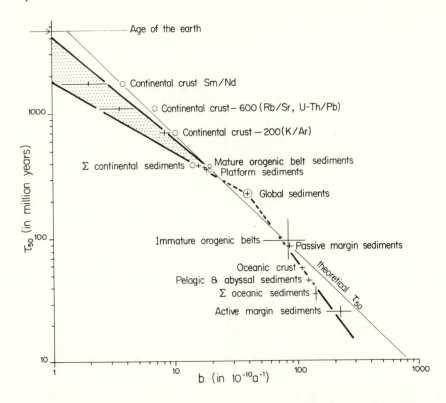

Figure 4 Plot of the observed half-mass and half-area ages (τ_{50}) and of recycling rates (b) for major global tectonic realms. Crosses represent solutions for the "preferred" model of continental growth, where it is assumed that continents started to grow (or started to be preserved) at ~4000 Ma ago. This was followed, ~3000–2000 Ma ago, by a fast growth rate, and subsequently the continents have been of a near present-day size. The circles give results for the steady-state (continents of present-day size) alternative. The two limiting solutions are identical at $b \geq 40 \times 10^{-10}a^{-1}$. For definition of b see Figure 1. Continental crust -200, -600 represent solutions for the bottom and the top enveloping curves in Figure 3. The Sm/Nd curve is not shown here. Reproduced from Veizer and Jansen [53].

phenomena and, following uplift, mountain ranges are eroded back into the oceans.

The above discussion may imply a notion of a steady-state Earth of infinite age. This, of course, is not the case. Any population, regardless of whether it is living or geological, may recycle and evolve simultaneously. Thus the hominid population branched out (was "born") about four million years ago and has been increasing ever since through recycling of numerous generations. In contrast, the population of dinosaurs became extinct about 65 million years ago despite generational procreation. In the case of the Earth, it is widely accepted that our planet was "born" ≈ 4.5 billion years ago, and during subsequent times generated, through internal differentiation, its present-day geological framework of continental and oceanic domains, as well as its hydrosphere, atmosphere, and biosphere. The evolution of the Earth is characterized therefore by a combination of recurring (cyclic) and superimposed unidirectional phenomena. The recurring, shorter term, phenomena leave behind only residual records, and the long-term trends, as they appear to us, are usually a compounded picture of such residual records. For example, the present-day continental crust is a result of agglomeration of remnants from recurring episodes of orogenesis. To understand this long-term evolution, it is essential to appreciate the time scales of the repetitive phenomena and their probabilities of preservation over geologic time. It is evident that tectonic realms of the oceanic domain are preserved only in a fragmentary fashion in geologic segments older than ≈ 300 Ma. The direct study of the oceanic tectonic domain must therefore be confined to this relatively recent time interval. Its preceding evolution can be reconstructed only from attributes of some derivative property, the latter measurable over a longer time span and having time resolution of $\leqslant 10^7$ years. In contrast, the evolution of the slowly recycling continental tectonic domain is discernible only on protracted time scales of 10^8–10^9 years, regardless of whether it is studied through direct or derivative observations. The ideal bookkeeping technique should therefore cover the bulk of the terrestrial life-span and have signal resolution on a variety of time scales. The isotopic properties of coeval sea water have a potential to serve as such a tool.

The Hydrosphere-Atmosphere Hierarchy of Populations

The present-day oceans and atmosphere have masses of 10^{24} and 10^{21} g, respectively [30, 44] and their compositions and evolutions have been treated comprehensively [18, 25, 26, and 56]. The details of the age distributions of water and atmospheric masses are only partially known [see 8], but there is little doubt that these age distributions conform with the exponential or power law pattern advocated above [30]. The yearly flux of river water into the oceans is $\approx 0.3 \times 10^{20}$ g. Thus the evaporation-precipitation-runoff cycle is capable of recycling the

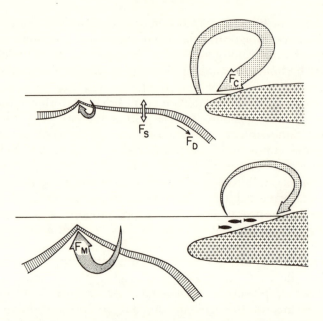

Figure 5 Schematic representation of fluxes controlling the composition of sea water. F_C "continental" river discharge; F_M hydrothermal and low-temperature "mantle" flux, involving circulation of heated sea water through ocean floor basalts; F_S sedimentary flux involving deposition of sediments and their interaction with sea water; and F_D flux of sediments and water subducted into mantle.

whole ocean in $\approx$40,000 years, and this is designated as the *mean residence time* τ_{res}. In most instances, $\tau_{max} \geqslant \tau_{res} \geqslant \tau_{50}$. Consequently, the probable half-life for the entire population of oceanic H_2O is about ten thousand years. The residence times for chemical and isotopic species dissolved in sea water range from $\leqslant 10^2$ to $\approx 10^8$ years [see 25], and for the smaller populations of atmospheric constituents they are about 10^{-2}–10^7 years [30]. The corresponding half-lives for all the above dissolved species and atmospheric constituents are therefore likely to be somewhat shorter than their residence times.

CONTINENTAL DOMAIN: DERIVATIVE SIGNAL IN SEA WATER

The chemical and isotopic composition of sea water reflects average global causes and not local phenomena. For this reason, a time record of its evolution is desirable information. Unfortunately, there are no

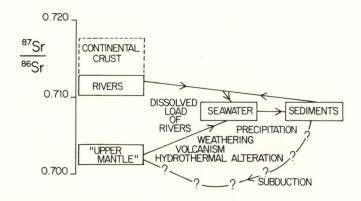

Figure 6 Schematic presentation of Sr isotopic surficial cycle. Reproduced from Wadleigh [55].

unequivocal samples of ancient sea water (or of air) preserved in the rock record. Its properties therefore can be deciphered only through derivative signatures inscribed in ancient marine sediments.

The present-day fluxes controlling the composition of sea water are:
(a) continental river discharge F_C;
(b) interaction between sea water and oceanic basalts ("mantle" flux), F_M;
(c) efflux via, and interaction with, sediments F_S; and possibly
(d) subduction of sediments and trapped water in subduction zones F_D (see Fig. 5).

As already stated, in terms of water transfer the river flux is capable of recycling one ocean volume in $\approx 4 \times 10^4$ years, whereas the "mantle" flux requires $\approx 2 \times 10^7$ years [15]. The estimates for the net unidirectional sedimentary (F_S) and subduction (F_D) fluxes are uncertain and probably small compared to those of the river and "mantle" fluxes. Taking into account that sea water recycled through oceanic basalts has the concentration of solutes $\approx 10^3$ times higher than river water, the "mantle" flux has, for some elements, the potential to match the magnitude of the river flux. The change in relative intensities of F_C:F_M fluxes, that is, in relative importance of the continental vs oceanic tectonic domain, results in secular (time-related) evolution of chemical and isotopic composition of sea water. This evolution is, in turn, recorded in (bio)chemical sediments, such as carbonates. Consequently, the secular trends in the latter provide a direct weighted record of past perturbations. Tracers with differing signatures for these two fluxes are particularly suitable for such an approach. For example, the

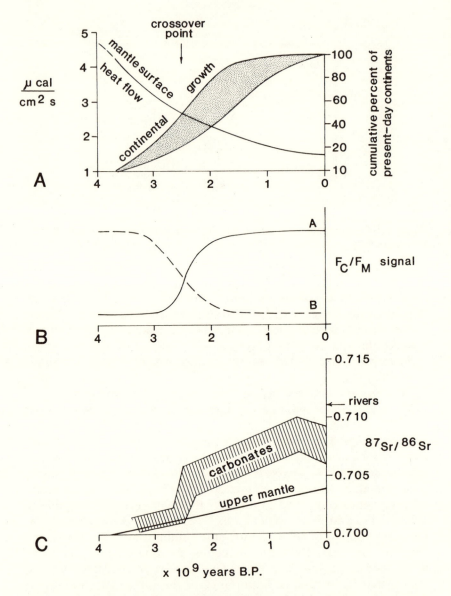

Figure 7 A Schematic presentation of mantle surface heat flow and of areal
continental growth during geologic history.
B The expected F_C/F_M signal in sea water composition. This signal should be of
type A if the transition results in a gain, and of type B if in a loss of an entity.
Note that this general shape of the signal will evolve regardless of the details of

average Sr isotopic composition of river water is 0.711, whereas that of oceanic basalts is 0.703 (Fig. 6). The sea-water value of 0.709 therefore demands that the river flux of Sr outweighs its "mantle" counterpart by ≈4:1. In the geological past, sea water $^{87}Sr/^{86}Sr$ fluctuated between the above end-members. The river flux was probably non-linearly proportional to the size of the coeval continents (Fig. 7A). In contrast, the intensity of water circulation through ocean-floor basalts was probably proportional to mantle heat production and its dissipation. Thus the buffering of ocean chemistry when the Earth was young, with high mantle heat generation and small continents, should have been mostly by basalts (F_M), whereas during the later periods – characterized by large continents and less mantle heat production – the dominant controlling flux was the river flux (F_C). The resulting Sr isotopic signal should therefore approximate that of A in Figure 7B. Indeed, the first order (10^8–10^9 years) $^{87}Sr/^{86}Sr$ secular trend, as observed in sedimentary carbonates, approximates the proposed signal (Fig. 7C), with the transition from a "mantle" to a "river" buffered ocean apparently accomplished ≈2500 Ma ago.

The above observations, although open to alternative interpretations, are in good agreement with the available geologic record [see 59]. The oldest continental-type rocks are known from the Isua area of Greenland, and their age is ≈3800 Ma [35]. The only older terrestrial materials are cores of some detrital mineral grains such as zircons ($ZrSiO_4$) from sediments in Australia, with ≈4200 Ma ages [16]. Following the "birth" of the continental domain (population), its growth increased at first exponentially, but subsequently slowed down and approached a near steady-state situation at ≈1800 Ma ago. This growth pattern again resembles a population growth which approaches the carrying capacity of the system, as in living examples [cf. 14]. What is being argued at this stage is the relative importance of the generation of new pristine continental crust from the mantle, as opposed to the destruction of the existing crust, as a principal cause of this apparent continental growth pattern. If the rate of recycling of

---

the two exponential (or other type) curves of opposing slopes. For radiogenic isotopes (i.e., Sr) the flat parts of the curves will be modified by radioactive decay.

c Schematic $^{87}Sr/^{86}Sr$ variations in sea water during geologic history based on the data in reference 52. Note the departure of the "sea water" curve from mantle values at ~2500 Ma ago. Reproduced from Veizer [50].

continental crust through mantle were diminishing in the course of geologic evolution, the proposed continental growth pattern could simply be a consequence of a preservation probability. It is likely that both causes were involved. Nevertheless, the lack of pre-3000 Ma detrital components in sediments of all ages [49, 33] argues for a general scarcity of old continental crust and thus for accelerated growth as a major reason for continental build-up during the early, Archean, interval of terrestrial evolution.

OCEANIC DOMAIN: DERIVATIVE SIGNAL IN SEA WATER

The tectonic populations of the continental domain operate on time scales of 10^8-10^9 years, and the first-order Sr isotopic curve therefore reflects the relative intensity of the river flux only on this time scale. Ever since continents and rivers became large enough to dominate the surficial cycle of Sr, sea-water isotopic composition of this strontium never dropped back to mantle values. The "mantle" flux, which reflects the operation of the faster, but smaller, oceanic domain, operates on time scales of 10^7 years. This causes second-order oscillations of 10^7 years wavelength superimposed on the first-order, 10^8-10^9 years, trend. Such oscillations have been detected as far back as the Archean, but the existing geochronology does not provide the desired time resolution on a routine basis until the appearance of shell-secreting organisms at the beginning of the Phanerozoic, some ≈ 570 Ma ago (Fig. 8). That these variations are indeed a reflection of the operation of the oceanic tectonic domain is confirmed by a correlation of Sr isotopic composition of sea water with "sea-level" stands (Fig. 9). The latter are relative heights of paleosea-levels in relation to the present-day strandlines. Theoretically, a higher rate of heat dissipation from the mantle causes more volcanism and thus faster generation and spreading of the basaltic oceanic crust. The latter, in turn, displace sea water and cause marine transgressions over continental margins [23]. The "sea-level" curve for the Phanerozoic should therefore be a direct reflection of the intensity of sea-floor spreading in the geologic past. Higher spreading rates also demand more efficient cooling of the newly generated oceanic crust by more pervasive convective circulation of sea water (Fig. 5). This factor introduces non-radiogenic basaltic $^{87}Sr/^{86}Sr$ into coeval oceans. However, contrary to expectations, the high "sea-level" stands are associated with influx of radiogenic continental river-born Sr ($^{87}Sr/^{86}Sr \approx 0.711$). The observed correlation of $^{87}Sr/^{86}Sr$ with "sea-level" shows, therefore, that at times of fast spreading the

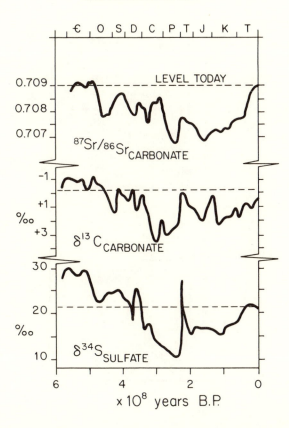

Figure 8 Sea water isotopic age curves for the Phanerozoic (latest 570 million years) as recorded (a) in carbonates and in fossil apatites for Sr isotopes; (b) in carbonates for carbon isotopes; and (c) in evaporite sulfates for sulfur isotopes. Modified from Holser [27] and Veizer [50].

denudation of continents increases to such a degree that "continental" Sr overwhelms any hydrothermal increase in addition of "mantle" Sr, the present-day balance being ≈4:1. In other words, fast sea-floor spreading leads to mountain building and increased erosion, as indicated already from the consideration of recycling rates. The Cretaceous (≈140-65 Ma BP) ocean appears to have been a major exception from the general Phanerozoic steady-state. Its non-radiogenic nature indicates very high influx of basalt-derived Sr. A massive intraplate volcanic activity, such as that from a multitude of Hawaiian-

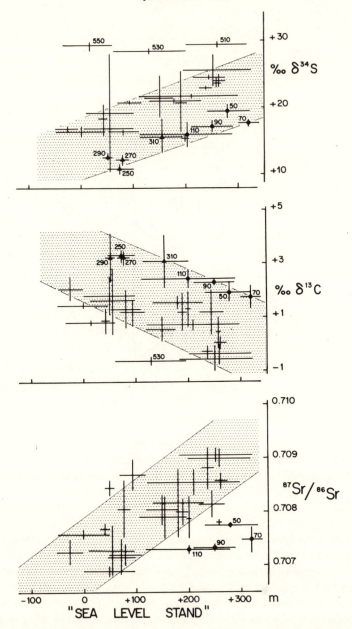

Figure 9 Scatter diagrams of sea water $\delta^{34}S$, $\delta^{13}C$, and $^{87}Sr/^{86}Sr$ versus "sea level" stands during the Phanerozoic. The isotopic data are taken from the latest compilation of the curves in Holser [27] and "sea-level" stands from Vail

type island chains [see 42], may have been the cause. Such volcanism does not necessarily result in tectonic compression and mountain building at the ocean-continent interface and a consequent influx of radiogenic ^{87}Sr.

Population Hierarchy of Living Systems

In terms of organic carbon (C_{org}) the mass of the extant global biosphere is perhaps $\approx 10^{18}$–10^{19} g, with constituent populations being orders of magnitude smaller [4, 5]. The life-spans of these populations are from centuries to hours, and their corresponding half-lives (τ_{50}) are therefore in the range of 10^2-10^{-3} years. The general compatibility of internal age distributions in living systems with the pattern described in Figure 1 is well established [see 30, 11, 45, 31, 36, 21, 48, and 28]. From the geological point of view, it is important to ascertain when the terrestrial biosphere first appeared and what was its subsequent evolution towards the present-day steady-state. The evolutionary aspects are treated in Schopf [43] and by McLaren in this volume. This contribution will deal only with the origin and the size of the terrestrial biosphere.

THE ANTIQUITY OF LIFE

The oldest direct evidence for the existence of terrestrial life is the discovery of stromatolites (Fig. 10), laminated organosedimentary structures produced by microbial biocenoses, in the ≈ 3500 Ma-old Warrawoona Group of Western Australia, and of filamentous prokaryotes (Fig. 11) from a somewhat uncertain location in the same sedimentary sequence [cf. 43, 20]. It appears that life on Earth spans almost the entire time interval represented by the preserved rock record. The oldest known rocks, the ≈ 3800 Ma-old Isua metasediments, contain only disputed fossil-like artefacts [37]. These sediments are, however, strongly thermally altered and even a negative outcome of the present discussion should not be regarded as evidence for the

---

and Mitchum [47]. The crossed bars represent the boundaries for a square which encompasses all variations for a given ~20 Ma interval. The bars therefore signify the maximal limits of the observed variations, and their intersect signifies the midpoint (not average) of these variations. Numbers 70, 50, etc mark the 70 ± 10, 50 ± 10 Ma ages for those time intervals which depart from the norm. Reproduced from Veizer [51].

Figure 10 Stratiform stromatolitic laminae from strata of the 3500 Ma-old Warrawoona Group in the North Pole Dome region of northwestern Australia. Reproduced from Schopf [43].

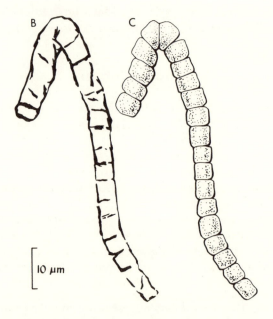

Figure 11 Unbranched, septate, filamentous prokaryote in schematic reconstruction from stromatolitic black chert of the 3500 Ma-old Warrawoona Group. Reproduced from Schopf [43].

absence of life. The great antiquity of life is supported also by arguments based on the inventory of terrestrial carbon, the fundamental building block of living systems.

The global inventory of carbon in the overall sedimentary, or exogenic, system of the Earth (comprising sediments, oceans, atmosphere, and life) is $\approx 10^{22}$ g, with the surficial exchange reservoirs (atmospheric CO_2, dissolved marine bicarbonate, and extant biomass) accounting for only $\approx 10^{19}$ g of this amount [40, 41, 4, 5]. Consequently, the bulk of carbon resides in sediments either in its oxidized (carbonate rocks) or reduced (disseminated C_{org}) form, the present-day relative distribution between these two oxidation states being $\approx 80:20\%$. The transfer of carbon from a carbonate reservoir into C_{org} is achieved through weathering of sedimentary rocks, liberation of CO_2, and utilization of the latter in a photosynthetic reaction:

$$2 H_2O + CO_2 \rightarrow CH_2O + H_2O + O_2.$$

This results in generation of "new" organic substances (C_{org}) in the exchangeable reservoirs. The bulk of the biomass is reoxidized to CO_2 after death and is thus involved in rapid recycling on time scales of 10^{-3}–10^2 years. The fraction of C_{org} incorporated into sediments is $\leqslant 1\%$, and only this fraction is involved in the slow, 10^7–10^9 years long, rock cycle. Thus the sediment-water (atmosphere) interface represents – in analogy to mountain belts – an interface of a small but fast active cycle with a large and slow passive cycle.

The $\delta^{13}C$, an arbitrary measure of the ratio of $^{13}C/^{12}C$ isotopes relative to a given standard, for gases originating in the interior of the Earth, the terrestrial mantle, is $\approx -6\%_0$ [12]. In the absence of photosynthetic life, all surficial reservoirs would inherit a comparable $^{13}C/^{12}C$ ratio. On the other hand, photosynthesis is accompanied by a large isotopic fractionation, which, although somewhat variable, yields overall depletion of $\approx 25\%_0$ in ^{13}C in the generated organic matter. Assuming constant isotopic fractionation, the first photosynthetic life should have inherited $\delta^{13}C_{org}$ of about $-31\%_0$ (Fig. 12). With increasingly larger amounts of C_{org} stored in sediments, the $\delta^{13}C$ of the residual oceanic bicarbonate should become progressively heavier, reaching $\sim 0\%_0$ at a time when C_{org} attained $\sim 20\%$ of the total carbon pool [6]. The $\delta^{13}C$ of the coeval C_{org} should parallel this trend towards ^{13}C enrichment (Fig. 12). However, the observed first-order pattern resembles the present-day steady-state for the entire geologic record (Fig. 13). This indicates not only a great antiquity of life but also its

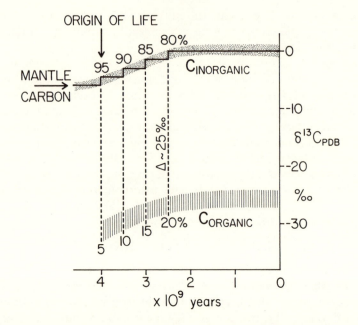

Figure 12 Schematic presentation of $\delta^{13}C$ evolution for a theoretical exogenic cycle of carbon. In this instance, it is assumed that life originated ~4000 Ma ago. The numbers 5, 10, 15, 20 are the percentage proportions of carbon which was sequestered into the reduced carbon reservoir (extant and buried C_{org}). $\delta^{13}C$ is derived as

$$\delta^{13}C \ (‰) = \left[\frac{(^{13}C/^{12}C) \text{ sample}}{(^{13}C/^{12}C) \text{ standard}} - 1 \right] \times 10^3.$$

$\delta^{34}S$ is derived in a similar manner. The standard for C is PDB (Pedee Belemnite), and for S it is CDM (Canyon Diablo Meteorite).

essential conservatism. Stated in simplified terms, a photosynthetic prokaryotic biosphere of ~1/2 [22] to near present-day size [41] must have evolved earlier than 3500 (3800) Ma ago, and subsequent development has been mostly that of recycling with ever-increasing biological complexity. These observations are perhaps unexpected but not entirely surprising. The limits on the size of the terrestrial biosphere, or the carrying capacity of the Earth, may be set, for example, by the supply of nutrients, such as phosphorus or nitrogen. This supply is a function of their release from rocks by weathering and thus is controlled by the rates of operation of the higher hierarchies of

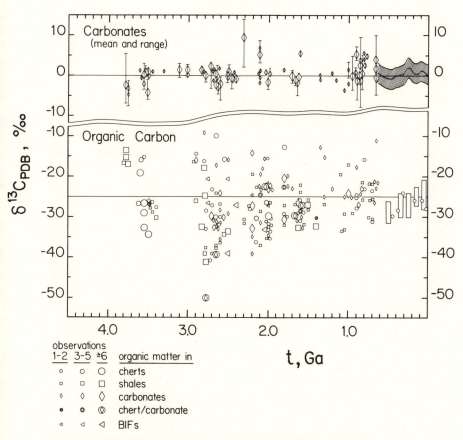

Figure 13 Isotopic composition of sedimentary carbon as a function of time. The approximate constancy of the carbonate record is in marked contrast with the scatter of data points for organic carbon (kerogen) that reflects both primary variations and the later effects of diagenesis and metamorphism. The gross average for sedimentary organics of all ages is approximately -25‰. The age trend for Phanerozoic carbonate values (mean and standard deviation) is according to Veizer et al. [54], and that for kerogens of the same age is according to Galimov et al. [17] and Welte et al. [58]. Bars imposed on the Phanerozoic kerogen record are standard deviations for selected age groups as reported by Degens [10]. Modified from Schidlowski et al. [41].

populations (solid earth and hydrospheric cycles). The emerging biosphere, because of the initial availability of nutrients, probably expanded exponentially, and rapidly reached the limits imposed by nutrient supply. Alternatively, the limits on the size of the terrestrial

C_{org} may have been set by the overall oxidation-reduction balance of the surficial Earth [26]. Once the limits were reached, further propagation of life was maintained mostly by internal cannibalism, as expressed in the rapid oxidation-reduction cycle of the extant biomass.

FATE OF BYPRODUCTS: THE CASE OF OXYGEN

The above scenario poses problems for the traditional view that evolution of the atmospheric oxidation state was controlled solely by advancing biological evolution. Although oxygen is not an inevitable byproduct of early photosynthesis and of C_{org} storage in sediments, the data are consistent with such an interpretation. The traditional view [see 9, 44] presupposed the gradual evolution of oxygen-producing photosynthetic biomass and, as a consequence, development of an oxygenic atmosphere-hydrosphere system, with a crossover point at $\sim 2100 \pm 300$ Ma ago. In the present scenario, the early sources of oxygen appear to have been plentiful [49], yet the sedimentological criteria indicate low oxidation state for the early hydrosphere-atmosphere system [cf. 57]. An effective sink (or sinks) is therefore required for the maintenance of low partial pressures of oxygen. On the early Earth such a sink could have been the previously advocated enhanced circulation of sea water through oceanic crust [49]. The present-day "black smokers," submarine hydrothermal vents which discharge modified, oxygen-depleted, hot sea water are an eloquent testimony to the potential effectiveness of this process. The scenario appears to be supported also by oxygen balance calculations. Since photosynthesis results in generation of two free oxygens for each C_{org} present either in the living biomass or buried in sediments, the total quantity of photosynthetic oxygen in the surficial terrestrial reservoirs on the surface should be twice that of C_{org}, that is, $\sim 3.2 \times 10^{22}$ g [40]. The known oxygen reserves in the hydrosphere-atmosphere system and in sedimentary products of iron and sulfur oxidation do not balance this production. It has been therefore proposed [1, 39] that the missing oxygen has been reinjected into the interior of the Earth, and the submarine hydrothermal systems (flux FM) are the prime candidates for such oxygen "scrubbers."

TECTONISM AND LIFE

Figure 13 demonstrates the essential stability of the biosphere on a time scale of $10^8 - 10^9$ years. Superimposed on this steady-state are second and higher order oscillations of $\leq 10^7$ years wavelengths. Because of time resolution problems, the second order oscillations in $\delta^{13}C$ have been deciphered for the Phanerozoic (Fig. 8), and the higher order variations for the Tertiary and Quaternary [see 46], intervals only. The observed 10^7 years oscillations again correlate with "sea-level" stands (Fig. 9), clearly identifying tectonic processes of the oceanic domain as a causative factor. In the present context, it is important to note that the higher the "sea-level," the lighter the $\delta^{13}C$ of sea water. This can be interpreted as an indication of a diminished sequestering of carbon into organic cycle; the latter preferentially subtracting the lighter (^{12}C) isotope. The precise geological meaning of this relationship has yet to be deciphered. Nevertheless, the observation clearly poses problems for models suggesting large creation and burial of biomass at the times of high "sea-level" and thus of large epicontinental seas [2, 3, 32]. It may also pose problems for models which propose second and higher order biomass control through supplies of nutrients, such as phosphorus [7]. Times of high erosional rates (high $^{87}Sr/^{86}Sr$) should also be the times of maximal nutrient supply by rivers to the oceans and not vice versa. Whatever the detailed scenario, closer inspection of $\delta^{13}C$ data (Fig. 9) indicates two limiting modes of operation of the carbon cycle. The "heavy" isotopic mode (full circles), characterized by more pronounced generation and burial of organic matter, has been dominant during the Late Paleozoic and Cretaceous periods. It may or may not be a coincidence that the two most important Phanerozoic mass extinctions, the terminal Paleozoic and Mesozoic ones [38], coincide with the terminations of this mode of ocean operation.

TECTONISM, LIFE, AND ATMOSPHERE

The negative correlation between $\delta^{34}S$ and $\delta^{13}C$ (Figs. 8, 9) [54] is a consequence of the coupling of sulfur and carbon surficial cycles [19]. The actual geological scenario is not well understood, but the chemical coupling can be written [27] as

$$8\,SO_4^{2-} + 2\,Fe_2O_3 + 8\,H_2O + 15\,C_{org} \rightleftharpoons 4\,FeS_2 + 15\,CO_2 + 16\,OH^-.$$

The net effect is the utilization of oxygen from SO_4^{2-} for oxidation of C_{org} into CO_2. The complementary reduction of sulfur, from SO_4^{2-} to H_2S, and ultimately pyrite, is bacterially mediated and results in an appreciable isotopic fractionation effect, with the light ^{32}S isotope concentrated in the reduced sulfur phase. In analogy with the carbon cycle, when geologic storage of reduced sulfur is large the remaining sulfate in the oceanic reservoir is heavy (depleted in ^{32}S). Because of the above sharing of oxygen, an increase in generation and burial of

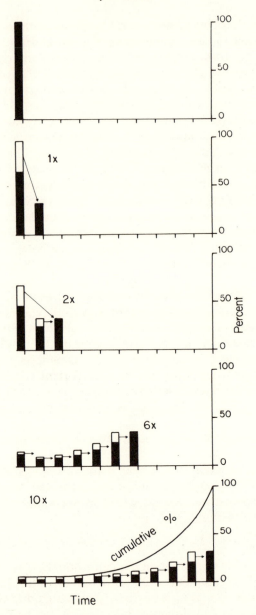

Figure 14 Schematic presentation of the present-day exponential (power law) cumulative age distribution of an entity generated via recycling in a steady-state population. The upper three cases may be indicative of age distributions in populations of galaxies and stars.

organic carbon results in enhanced oxidation of reduced sulfur and vice versa. This explains the observed negative correlation between $\delta^{13}C_{bicarbonate}$ and $\delta^{34}S_{sulfate}$ (Figs. 8, 9). As in the case of carbon, oceanic sulfur cycle operates again in two limiting modes, but with a reversed signature (see Fig. 9).

The correlations in Figure 9 demonstrate unequivocally that, on 10^7 years time scales, the solid Earth, hydrosphere, atmosphere, and biosphere act as a unified system. Consequently, the "living" and "inorganic" hierarchies of populations should not be envisioned as independent entities. Furthermore, the data show that the solid Earth and its internal tectonic cycles are the dominant causative hierarchy. Tectonism apparently controls "sea-level" stands, variations in chemical and isotopic composition of ocean water ($^{87}Sr/^{86}Sr_{carbonate}$), organic productivity and burial ($\delta^{13}C_{carbonate}$), and – via atmospheric linkage of C and S cycles – the $\delta^{34}S$ of marine sulfate. Although not yet understood in their precise geological scenarios, the data hint at a considerable buffering capacity and robustness of the terrestrial exogenic system in general, and of its life in particular. This buffering is a consequence of the linkage and interdependence of a multitude of diverse exogenic cycles.

Evidence for Design?

The preceding discussion indicates that, from the planetary point of view, the common thread to diverse physical phenomena is the concept of population dynamics. The populations discussed here span at least thirteen orders of magnitude in mass ($10^{13}-10^{26}$ grams). The age distributions of their constituent units conform to the exponential (power law) systematics typical of populations propagated by recycling, that is, through "birth" and "death" of the constituent units. Overall, annual recycling rates appear to have been several orders of magnitude ($\sim 10^{-6\pm4}$) smaller than the population size. In general, the life-spans (τ_{max}) and half-lifes (τ_{50}) of large populations appear to have been long and those of small populations short. The observed τ_{50}s ranged from 10^9-10^7 years for the geologic tectonic realms, to 10^8-10^2 years for the oceans, 10^7-10^{-2} years for the atmospheric constituents, and 10^2-10^{-3} years for the living systems. Viewed in this perspective, all populations are integral constituent parts of an all-encompassing unity. However, the concept of population dynamics does not require propagation of the present-day steady-state. The diverse populations are in a constant intercourse on a variety of spatial and temporal scales, with some populations in apparent steady-state, others growing or declining, and still others being born or dying. In this scenario, the grand design appears to have been a unity, with perpetual, statistically self-regulated, internal motions. It is beyond the scope of my experience to contemplate whether the concept of population dynamics can be extended to populations of microscopic and subatomic, as well as megascopic sizes (for example, galactic populations of stars, population of galaxies in the universe, "many universes"). The age distributions of stars and galaxies appear to be dominated today by old rather than young constituent units (Peebles, this volume). If this is in fact the case, the concept is not applicable to gigantic dimensions. Alternatively, our universe, with its constituent galaxies and stars, is in a nyouthful stage of a very long "life" cycle, and thus still populated by units born as an early response to its "big bang birth" (cf. Fig. 14).

The previously documented interdependency of inorganic and living worlds is consistent with the view that life is an important parameter in planetary evolution. However, in contrast to the Gaia hypothesis that life regulates the terrestrial exogenic system for its own benefit (Lovelock, this volume), I propose that life is constrained by limits imposed by the interdependent larger populations. In the first instance, the development of a suitable planetary system is controlled by vagaries of stellar evolution (Ovenden, this volume). Secondly, geological evidence from this planet clearly points out that tectonism controls the biosphere and not vice versa. Temporary departures from these controls are possible, but only on time scales shorter than those of the controlling larger populations. Nevertheless, I believe that within the broad limits erected by the dominant populations, life exerts substantial modulating influence and acts as a catalyst. This catalytic effect, resulting in acceleration of otherwise sluggish inorganic processes, is responsible for the maintenance, for example, of the atmospheric blanket of oxygen and thus for our own existence and well-being.

I acknowledge the financial support of the Natural Sciences and Engineering Research Council of Canada, and would like to thank J. Hayes for typing the manuscript, E. Hearn, W.F. Schmiedel, and A.M. Steele for drafting the figures, and D.J. McLaren and B.R. Rust for reviewing the manuscript. Figures 1 and 4 have been published by permission of the University of Chicago Press, Figures 7 and 8 by permission of the Elsevier Science Publishing Inc., Figure 9 by permission of the American Geophysical Union, and Figures 10, 11, and 12 by permission of Princeton University Press.

REFERENCES

1 Arrhenius, G. 1981. "Interaction of Ocean-Atmosphere with Planetary Interior." *Adv. Space Res.* 1:37–48.

2 Arthur, M.A., W.E. Dean, and S.O. Schlanger. 1985. "Variations in the Global Carbon Cycle during the Cretaceous Related to Climate, Volcanism, and Changes in Atmospheric CO_2." In E.T. Sundquist and W.S. Broecker. 504–29.

3 Berner, R.A., A.C. Lasaga, and R.M. Garrels, 1983. "The Carbonate-Silicate Geochemical Cycle and its Effect on Atmospheric Carbon Dioxide over the Past 100 Million Years." *Amer. J. Sci.* 283:641–83.

4 Bolin, B., ed. 1981. *Carbon Cycle Modeling.* New York: Wiley.

5 Bolin, B., and R.B. Cook, eds. 1983. *The Major Biogeochemical Cycles and Their Interactions.* New York: Wiley.

6 Broecker, W.S. 1970. "A Boundary Condition on the Evolution of Atmospheric Oxygen." *J. Geophys. Res.* 75:3553–7.

7 Broecker, W.S. 1982. "Ocean Chemistry during Glacial Times." *Geochim. Cosmochim. Acta* 46:1689–1705.

8 Broecker, W.S., and T.H. Peng. 1982. *Tracers in the Sea.* Palisades, N.Y.: Eldigio Press, Columbia University.

9 Cloud, P.E. 1976. "Major Features of Crustal Evolution." *Geol. Soc. S. Afr. Trans.* vol. 79. Annexure.

10 Degens, E.T. 1969. "Biogeochemistry of Stable Carbon Isotopes." In *Organic Geochemistry.* Ed. G. Eglinton and M.T.J. Murphy. New York: Springer. 304–28.

11 Deevey, E.S. 1947. "Life Tables for Natural Populations of Animals." *Quart. Rev. Biol.* 22:283–314.

12 DesMarais, D.J., and J.G. Moore. 1984. "Carbon and Its Isotopes in Mid-Oceanic Basaltic Glasses." *Earth Planet. Sci. Letters* 69:43–57.

13 Dietz, R.S. 1961. "Continent and Ocean Basin Evolution by Spreading of the Sea Floor." *Nature* 190:854–57.

14 Dodd, J.R., and R.J. Stanton. 1981. *Paleoecology, Concepts and Applications.* New York: Wiley.

15 Edmond, J.M., and K. Von Damm. 1983. "Hot Springs on the Ocean Floor." *Scientific American* 248:78–93.

16 Froude, D.O., T.R. Ireland, P.D. Kinny, I.S. Williams, and W. Compston. 1983. "Ion Microprobe Identification of 4100 to 4200 Ma-old Terrestrial Zircons." *Nature* 304:616–18.

17 Galimov, E.M., A.A. Migdisov, and A.B. Ronov. 1975. "Variation in the Isotopic Composition of Carbonate and Organic Carbon in Sedimentary Rocks during the Earth's History." *Geochem. Internat.* 12:1–19.

18 Garrels, R.M., and F.T. Mackenzie. 1971. *Evolution of Sedimentary Rocks.* New York: Norton.

19 Garrels, R.M., and E.A. Perry. 1974. "Cycling of Carbon, Sulfur, and Oxygen through Geologic Time." In *The Sea.* Ed. E.D. Goldberg. New York: Wiley-Interscience. 303–36.

20 Groves, D.I., J.S.R. Dunlop, and R. Buick. 1981. "An Early Habitat of Life." *Scientific American* 245:64–73.

21 Hallam, A. 1972. "Models Involving Population Dynamics." In *Models in Paleobiology*. Ed. T.J.M. Schopf. San Francisco: Freeman, Cooper, and Co. 62–80.

22 Hayes, J.M. 1983. "Geochemical Evidence Bearing on the Origin of Aerobiosis, a Speculative Hypothesis." In J.W. Schopf. 291–301.

23 Hays, J.D., and W.C. Pitman. 1973. "Lithospheric Plate Motion, Sea Level Changes and Climatic and Ecological Consequences." *Nature* 246:18–22.

24 Hess, H.H. 1962. "History of Ocean Basins." In *Petrologic Studies*. Ed. A.E.J. Engel, H.L. James, and B.F. Leonard. Geological Society of America. 599–620.

25 Holland, H.D. 1978. *The Chemistry of the Atmosphere and Oceans*. New York: Wiley-Interscience.

26 Holland, H.D. 1984. *The Chemical Evolution of the Atmosphere and Ocean*. Princeton: Princeton University Press.

27 Holser, W.T. 1984. "Gradual and Abrupt Shifts in Ocean Chemistry during Phanerozoic Time." In *Patterns of Change in Earth Evolution*. Ed. H.D. Holland and A.J. Trendal. Berlin: Springer. 123–43.

28 Hutchinson, G.E. 1978. *An Introduction to Population Ecology*. New Haven: Yale University Press.

29 LePichon, X., J. Francheteau, and J. Bonnin. 1973. *Plate Tectonics*. Amsterdam: Elsevier.

30 Lerman, A. 1979. *Geochemical Processes*. New York: Wiley-Interscience.

31 MacArthur, R.H., and J.H. Connel. 1966. *The Biology of Populations*. New York: Wiley.

32 Mackenzie, F.T., and J.D. Pigott. 1981. "Tectonic Control of Phanerozoic Sedimentary Rock Cycling." *J. Geol. Soc.* 138:183–96.

33 Miller, R.G., and R.K. O'Nions. 1985. "Source of Precambrian Chemical and Clastic Sediments." *Nature* 314:325–30.

34 Morgan, W.J. 1968. "Rises, Trenches, Great Faults, and Crustal Blocks." *J. Geophys. Res.* 73:1959–82.

35 Moorbath, S. 1977. "The Oldest Rocks and the Growth of Continents." *Scientific American* 236:92–104.

36 Odum, E.P. 1971. *Fundamentals of Ecology*. 3rd ed. Philadelphia: W.B. Saunders.

37 Pflug, H.D. 1978. "Yeast-like Microfossils Detected in the Oldest Sediments of the Earth." *Naturwissenschaften* 65:611–15.

38 Raup, D.M., and J.J. Sepkoski. 1984. "Periodicity of Extinctions in the Geologic Past." *Proc. Natl. Acad. Sci. U.S.A.* 81:801–5.

39 Ronov, A.B. 1982. "Global Balance of Carbon in the Post-Algonkian." *Geokhimija* 7:920–32 (in Russian).

40 Schidlowski, M., and C.E. Junge. 1981. "Coupling among the Terrestrial Sulfur, Carbon and Oxygen Cycles: Numerical Modeling Based on Revised Phanerozoic Carbon Isotope Record." *Geochim. Cosmochim. Acta* 45:589–94.

41 Schidlowski, M., J.M. Hayes, and I.R. Kaplan. 1983. "Isotopic Inferences of Ancient Biochemistries: Carbon, Sulfur, Hydrogen, and Nitrogen." In J.W. Schopf. 149–86.

42 Schlanger, S.O., H.C. Jenkyns, and I. Premoli-Silva. 1981. "Volcanism and Vertical Tectonics in the Pacific Basin Related to Global Cretaceous Transgressions." *Earth Planet. Sci. Letters* 52:435–49.

43 Schopf, J.W., ed. 1983. *The Earth's Earliest Biosphere: Its Origin and Evolution.* Princeton: Princeton University Press.

44 Scientific American. 1983. "The Dynamic Earth." *Scientific American* 249, 3:46–189.

45 Slobodkin, C.B. 1962. *Growth and Regulation of Animal Populations.* New York: Holt, Rinehart, and Winston.

46 Sundquist, E.T., and W.S. Broecker, eds. 1985. *The Carbon Cycle and Atmospheric CO_2: Natural Variations Archean to Present.* Geophysical Monograph Series 32. Washington: American Geophysical Union.

47 Vail, P.R., and R.M. Mitchum. 1979. "Global Cycles of Relative Changes of Sea Level from Seismic Stratigraphy." *Am. Assoc. Petrol. Geol. Mem.* 29:469–72.

48 Valentine, J.W. 1973. *Evolutionary Paleoecology of the Marine Biosphere.* Englewood Cliffs: Prentice-Hall.

49 Veizer, J. 1983. "Geologic Evolution of the Archean-Early Proterozoic Earth." In J.W. Schopf. 240–59.

50 Veizer, J. 1984. "The Evolving Earth: Water Tales." *Precambrian Research* 25:5–12.

51 Veizer, J. 1985. "Carbonates and Ancient Oceans: Isotopic and Chemical Record on Time Scales of 10^7–10^9 Years." In E.T. Sundquist and W.S. Broecker.

52 Veizer, J., and W. Compston. 1976. "$^{87}Sr/^{86}Sr$ in Precambrian Carbonates as an Index of Crustal Evolution." *Geochim. Cosmochim. Acta* 40:905–15.

53 Veizer, J., and S.L. Jansen. 1985. "Basement and Sedimentary Recycling-2: Time Dimension to Global Tectonics." *J. Geol.* 93:625–43.

54 Veizer, J., W.T. Holser, and C.K. Wilgus. 1980. "Correlation of $^{13}C/^{12}C$ and $^{34}S/^{32}S$ Secular Variations." *Geochim. Cosmochim. Acta* 44:579–87.

55 Wadleigh, M.A. 1982. *Marine Geochemical Cycle of Strontium*. MSc thesis. University of Ottawa.

56 Walker, J.C.G. 1977. *Evolution of the Atmosphere*. New York: Macmillan.

57 Walker, J.C.G., C. Klein, M. Schidlowski, J.W. Schopf, D.J. Stevenson, and M.R. Walter. 1983. "Environmental Evolution of the Archean-Proterozoic Earth." In J.W. Schopf. 260–90.

58 Welte, D.H., W. Kalreuth, and J. Hoefs. 1975. "Age Trend in Carbon Isotopic Composition in Paleozoic Sediments." *Naturwissenschaften* 62:482–3.

59 Windley, B.F. 1984. *The Evolving Continents*. 2nd ed. New York: Wiley.

DIGBY J. McLAREN

An Anthropocentric View of the Universe: Evidence from Geology

I begin with an anecdote, a detached piece of history that I shall refer to again. It concerns Philip Henry Gosse, a Victorian zoologist and writer who, in 1857, once and for all resolved the dilemma of reconciling the facts of biology and geology with a belief in a sudden creation. He published his discovery in a book called *Omphalos, an Attempt to Untie the Geological Knot*. His solution involved two conflicting assumptions, the first of which required him as a member of a Calvinist sect known as the Plymouth Brethren, to believe that "in six days Jehovah made Heaven and Earth, the Sea, and all that in them is." And his second assumption was that all the facts of biology and geology are correct. I must emphasize that Gosse was a man of broad learning, well versed in many branches of science, a Fellow of the Royal Society, and correspondent of many of the leading scientists of his time. His argument was that as God could not create an imperfect being, at the moment of creation Adam had a navel, trees had tree-rings, and the teeth of a horse showed its age. Gosse examined the whole animal and plant world and concluded that it would be impossible to create any living thing at any stage in its life cycle that did not possess, at the moment it came into being, evidence of a past that had never existed. Gosse named time after creation "diachronic time," and time that appears to us as evidence for a long past "prochronic time" – time that existed in fact only in the mind of an ingenious Creator. We cannot follow Gosse further, but his importance rests in having given us the *reductio ad absurdum* of the literal creationist and, of course, his ideas cannot, by definition, be falsified, as he emphasizes that there can be no possible mark left in the record of the moment of creation.

One Universe or Many?

In considering the position of man in the universe, which is what I believe we are here to do, it is unlikely that any new theories will emerge concerning our origins. Indeed, the possibilities are summarized by Alan Batten in the Introduction to this volume. To the physicist, there have been only two kinds of universe: the classical mechanistic universe of Laplace in which the present was set in the machine at its beginning and in which the future must inexorably follow the past; and the universe of the uncertainty principle in which every event, after the fact, appears to have been highly improbable. P.C.W. Davies, for example, in *The Accidental Universe* repeatedly states that we are exceptionally fortunate to exist, and that our world is extremely unlikely on *a priori* grounds. Furthermore, as suggested by Everett, such a chance universe implies an infinite number of alternative universes, presumably in order that we can accept the one we find ourselves in without becoming too uneasy about our inherent improbability.

But the theme of probability, or improbability, underlies all our discussions. Physical theories of the origin and development of the universe have noted improbable coincidences at least since Eddington, and with the formulation of the anthropic principle, or principles, the search has broadened and we are told yet more frequently that we are exceptionally fortunate to exist. The inclusion of life in cosmological theory led Hoyle in 1954 to comment on the accident of the tuning of nuclear resonances in the synthesis of carbon and oxygen in stars. Without their being defined at precise values, carbon would be rare and life as we know it would be extremely unlikely. Hoyle has called this "a put-up job." Yet this is only a group of many coincidences that extend into the molecular world of chemistry, with much greater complexity, and on into the living cell and the evolution of complex organisms, which must be looked at in an entirely different manner. Prigogine and Stengers's transformation of physics, requiring interrelations connecting different forms of knowledge and diverse habits, is a significant example.

In a universe in which one accepts the *a priori* assumptions referred to above, then *ex post facto* probability makes every event of the past look highly improbable. If, however, we leave consideration of the relatively simple laws that determine how the very large and the very small may behave, and look back in time from a vantage point at the meso-scale of our own experience, we seem to see patterns, lineages, and linkages in the physical and biological history of the Earth. We can observe the formation of a system which progressed through chemical and biological evolution to the establishment of a controlled environment in which life has existed for up to 4 Ga. Lovelock's Gaia is as highly tuned as Hoyle's nuclear resonances. In 1965 Pantin emphasized the idea of emergence in this development: "The different grades of system, atoms, molecules, organisms and so on, display empirically emergent properties, so that fresh assumptions have to be made in hypotheses representing them over and above those needed for the lower-order systems from which they are built." Prigogine and Stengers have given a more formal structure to such observations and suggest that theories are emerging that make numerous aspects of the world more intelligible. Thenceforth the perspective cannot be that of a universal theory deducible from an exact definition of elementary units. We must include, in their terminology, "the global behaviour of the collective."[1]

The Planet Earth

When we consider our own position in the cosmological model that we have constructed, we find ourselves on a planet that is part of the solar system, whose age is about a third of the age of creation. This is our home and this is the platform from which our observations are made, and we are peculiarly well situated to examine in some detail the history of the last 4 Ga or so. This comes about because our planet, unique in the solar system, is an engine which stores its own history by a system which preserves evidence of physical and chemical changes over time in sedimentary and crystalline rocks. And we find that there is also a record of living processes on the planet nearly as old as the earliest rocks. We can thus witness our own development. At this point I would make a plea for empiricism, observational evidence versus *a priori* speculation, for we must realize that the only evidence we have concerning the age and development of life in the universe is in this imperfect record of traces, replicas, and chemical signatures of organisms that have constituted an unbroken biosphere for about four thousand million years.

The history of this restless planet has been pieced together following a few simple empirical laws for the most part set out by Hutton two hundred years ago. Yet geology offers a remarkably sensitive tool for reconstructing the past – a relative time-scale based on the positional relationships of rock and mineral bodies which allows us to judge what was happening simultaneously at many places on the crust. The scale is calibrated approximately in measured time by systems of dating largely based on decay of radioactive isotopes. From this long record, I should like to summarize the major changes in life through time, and then to consider the geological processeses that continuously changed the Earth, thereby forcing life to change in that endless process we call evolution. I propose to pause only on biological and geological matters less well known to a general audience, such as the early development of life up to the first major radiation of the Metazoa; recent ideas on the forces producing major environmental changes to the Earth's surface; the resulting crisis points in the biosphere and the phenomenon of major extinctions and their possible causes.

Biological Events in Time

This is not the place, nor do I have the capacity, to consider the origin of life. Any views must be highly speculative, even though the steps along the way may be reasonably well defined. One should note, however, that the reductionist approach shows that its appearance is statistically extremely improbable. Prigogine has suggested a way out of the dilemma in the theory of dissipative structures. Chance and necessity are no longer in conflict. For him "the most important consideration is that life no longer seems to be a precarious miracle, a struggle against a universe which rejects it ... [T]he living being is not a strange product of chance, nor the improbable winner in a huge lottery."

The crust of the Earth is believed to have formed some 4.2 Ga ago. The earliest rocks known form the Isua Complex of Southwestern Greenland and are dated at about 3.8 Ga (Fig. 1). They consist of metamorphosed volcanic and sedimentary rocks for which there is no known basement. The earliest fossils yet reported occur in the Warrawoona Group of Western Australia at about 3.5 Ga, and have been discussed by Schopf et al., who have given us the best overall view yet published of the early development of life. The fossils consist of silicified stromatolites containing micro-organisms which are judged to consist entirely of a prokaryotic biota, which probably includes both heterotrophic and autotrophic forms, some or all anaerobic. These organisms are already of a complexity and diversity that suggests to Cloud and Schopf that life may have started on Earth substantially earlier. Furthermore, carbon isotope ratios in sedimentary organic matter 3.5 Ga old and younger have values consistent with the

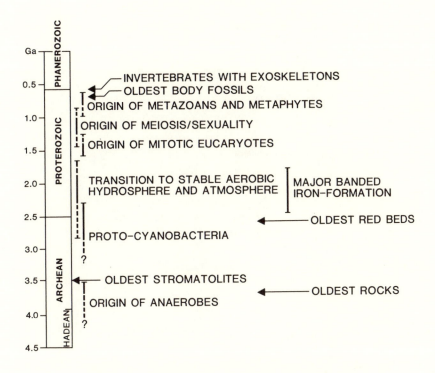

Figure 1 Major biological events in geological time

existence of autotrophs. The dominant life form by the late Archean may have been cyanobacteria (the so-called blue-green algae), which produces free oxygen in the ocean, although this process could have begun earlier.

At about 2.5 Ga the first red beds in the geological record suggest the beginning of higher levels of free oxygen in the atmosphere, coinciding approximately with the start of the major phase of deposition of the banded iron-formation, which took place during the transition to a stable aerobic atmosphere. As free oxygen increased there were changes in life involving the evolution of systems of protection against oxygen and systems of metabolic and biosynthetic oxygen use. Five principal categories of different morphologies are recognized, all prokaryote, and although actual fossil evidence remains rare, stromatolites continue abundant.

Throughout this long period of time, the major mode of evolution remained at the intracellular and biochemical level. Chapter one in life

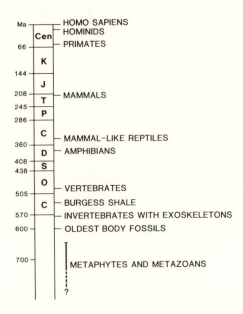

Figure 2 Biological events in the Phanerozoic

history ended with the origin of unicellular mitotic eukaryotes about 1.5 Ga, to be followed by the development of meiosis or sexuality, thus opening the way to organismal and morphological evolution.

The time of origin of multicellular animals and plants remains uncertain. Trace fossils from about 1 Ga of possible trails and feeding burrows, at first controversial, gradually increase together with signs of multicellular algae. Undoubted body-fossils, however, are unknown before 700 Ma. They are of considerable interest and Seilacher has suggested that forms of the famous Ediacara fauna, now known from four continents, although superficially similar to existing invertebrate phyla, may represent a distinct evolutionary trial that flourished until about 600 Ma, and then died out leaving no descendants. At about the same time there was a crash in the late Proterozoic algal micro-fossils to be followed by a new radiation in the early Cambrian.

So we come to the beginning of the Phanerozoic – revealed life, represented by a sudden development of many phyla of metazoans with skeletal structures, and therefore easily fossilized (Fig. 2). From this moment, about 570 Ma ago, there was a huge jump in our capacity to read the record of life which corresponded with a major radiation into most of the known phyla of animals within the Cambrian Period. We cannot trace the whole story through the record, but one fortunate chance has given us a better understanding of early life than we might have had. This was the freak preservation of a complete fauna of Middle Cambrian age including forms without skeletons in the Burgess Shale above Emerald Lake near Kicking Horse Pass in the Rockies of British Columbia. Whittington and his team have recently redescribed this remarkable development of organisms representing about ten of the known animal phyla, including the chordates. Two thirds of the taxa are soft-bodied, and therefore normally never fossilized. One sixth of the 119 genera cannot be assigned to a known phylum, and it has been suggested that as many as 15 unknown phyla might be present. The remaining events in the Phanerozoic shown on Figure 2 concern the stepping stones in the emergence of man. Chordates are known in the Burgess Shale and the first vertebrate remains found so far are from the Lower Ordovician. And so on, with decreasing taxonomic, rank to the primates and man.[2]

The Unquiet Earth

The Phanerozoic represents the last 15 per cent of time since the earliest rocks formed and contains the evidence for the major part of metazoan evolution. Orthodox geological and evolutionary theory have leaned towards gradualism, although during the last twenty to thirty years it has had to accept evidence of forces at work which would seem to suggest that the Earth's history has been very much more episodic than had been imagined. Today's "neo-uniformitarians" still maintain that the physical events of geology are always slow acting, and that, although speciation may be punctuated, all changes can be explained by orthodox evolutionary theory. Does geological evidence support such a conclusion? We now understand that there are three major causes of environmental change that affect the crust and its biosphere: (1) internal changes in the Earth, in core or mantle, of which the dominant are plate movements leading to a wide variety of effects, including volcanism, mountain building, longitudinal and latitudinal movements of continental plates, climatic change, changes in ocean volume and circulation, and changes in continentality, both in number and hypsographic character; (2) changes in the Earth's orbit, the sun's radiation, and the position of the solar system within the galaxy; and (3) astronomical events leading to large-body impact (asteroids and comets) and nearby supernovae (see Fig. 3).

IMMEDIATE CAUSES		ULTIMATE CAUSES		
		MANTLE CONVECTION	SOLAR AND ORBITAL CHANGES	IMPACTS AND SUPERNOVAE
SEA–LEVEL CHANGES		X		
CLIMATIC CHANGES	LONG TERM	X	X	
	SHORT TERM		X	X
RADIATION INTENSITY CHANGES	LONG TERM		X	
	SHORT TERM	X	X	X
POISONING		X		X
SHOCK EFFECTS				X

GSC

Figure 3 Causes of extinctions: some possible immediate causes that could have been initiated by one or more ultimate causes (McLaren 1983, cited in n 3)

These ultimate causes may bring about a variety of slow or quick acting effects that influence the surface of the Earth and life, including sea-level changes leading to changes in species area, short and long-term climatic changes (such as seasonality, glacial epochs, variation in atmospheric CO_2, and short-term temperature changes), variation in solar radiation intensity, magnetic reversals, ocean poisoning by overturn and mixing of anoxic waters, shock effects (such as including tsunami), high turbidity, atmospheric dust and water particles, darkness, and nitrogen oxides. Because there is not necessarily a one-to-one correspondence between immediate effects and any one of the three ultimate causes, these has been much controversy. A common problem in empirical science lies in the fact that confirming or falsifying a sub-hypothesis may have little or no bearing on the main theory.

Such ideas have only relatively recently become respectable, and I should like to examine the last-named ultimate cause in a little more detail. The evidence is quickly summarized. Large-body impacts occur regularly. We have evidence of them in over 100 impact craters throughout the world, most of which have been reasonably well dated. The frequency of arrival of these bodies on Earth during the last 500 Ma is found to be consistent with the average rate of bombardment of the moon during the last 3.3 Ga. (For further discussion of the evidence see Shoemaker's recent papers.) A third line of evidence is presented by the discovery of asteroids in Earth-crossing orbit (Apollo objects and others) with a visible population (i.e., bodies larger than 0.5 km in diameter) of between 1200 and 2000, which is believed to remain relatively constant. Replacements from the asteroid belt and by de-gassed comets roughly balance the rate at which such objects are expelled from the solar system by orbital perturbation, or collected by the sun, the Earth, and other inner planets. Their calculated size and rate of arrival on Earth agrees with the lunar and Earth census data. Shoemaker has plotted arrival frequency against equivalent energy on impact. Twelve bodies with a diameter of about 0.5 km should arrive in a million years and, if on a continent, they will produce craters of about 10 km in diameter. Larger bodies of 10 km and more are known. The frequency of arrival of a 10 km body is about one every 50 Ma, or about 12 in the Phanerozoic. On a continent it would produce a crater of about 100 km, and we know of 3 craters of this size. The energy release from such a body is estimated at 10^{23} joules or 10^8 megatons. Such an energy release could result in an impact-winter of several months duration caused by stratospheric dust and ice crystals. If in the sea, the resulting ocean turnover and turbidity would be world-wide, and the tsunami hitting the coasts would be gigantic. The fireball would result in large amounts of NOx in the atmosphere causing world-wide and prolonged acid rain and a reduction in the ozone layer leading to increased UV-radiation over an extended period following the clearing of the dust.[3]

Extinction

Over much of the Phanerozoic, an ordered progression of life has certainly been the norm, but it has periodically been disrupted by environmental shock, induced by major forces described above, and throughout the whole of this history there have been accidents, crisis points. These were brought about by huge changes in the physical and chemical environment, some slow acting, others apparently fast, if not instantaneous. Some appear to have followed spaced rhythms or episodes, others seem to have occurred entirely by chance. In examining some examples from the Phanerozoic we may postulate that the pattern of this double system must also have obtained over the vast time span of the Precambrian. Figure 4 shows some of the larger

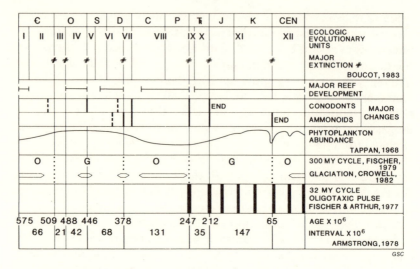

Figure 4 Some major extinctions plotted against various rythmic or episodic changes during the Phanerozoic (McLaren 1983, cited in n 3)

extinctions known, their age and spacing in time, and some associated phenomena.

I should like briefly to look at two of these extinctions. The Cretaceous-Tertiary (K-T) boundary is marked by a major extinction of the marine micro-fossil group, the foraminifera. Alvarez and others, by a serendipitous discovery, showed that the horizon of this extinction is marked by a geochemical anomaly in a thin clay layer that they suggest was caused by fallout from an explosion by a large body impacting the Earth. This anomaly is now known in four oceans from deep sea cores and in shallow marine and continental deposits over a large part of the globe. So much is fact, but controversy surrounds almost every other claim in regard to what became extinct and how quickly it all took place. Granted the extinction and the event, the impact theory is probably gaining ground, although the volcanic origin proponents are still in the field. The foram extinction is impressive and shows the pattern of a total wipe-out followed by a new radiation in essentially an empty environment, based, possibly, on a single surviving lineage.

In the 1960s I followed a major extinction in shallow marine benthos at the top of the Frasnian Stage of the Upper Devonian through western and northern Canada, and subsequently traced it across Europe, Siberia, and China to Australia. It appeared abrupt, and in

1970 I suggested a meteorite impact in the ocean, considering Dietz's demonstration of what might happen in such an event. Following an analysis of the ecology of the organisms that disappeared, I used the impact to produce a turbid ocean which would dispose of the largely filter-feeding organisms and their larvae. The biomass disappearance was huge, possibly over 90 per cent in some groups, and included all reef-building corals and their associates as well as many others, such as many trilobites, brachiopods, and fish. The rare, deep, cold water corals were unaffected. The succeeding Famennian Stage saw a radiation which resulted in the precursors of the shallow water Carboniferous coral fauna and the beginning of a new brachiopod radiation; trilobites continued, but never recovered their previous numbers or diversity. The postulate gained respectability when with Playford and others we found a geochemical anomaly associated with the extinction horizon in western Australia in 1984, and acceptance of the importance of the extinction is gaining ground. The cause must remain unproved. It is currently being studied all over the world.

The largest extinction recorded in Phanerozoic time was at the end Permian Period, a time of maximum continentality, high relief, lowest sea level, and severe extremes of heat and cold. Life suffered severely in the sea and on land throughout the latter part of the period. Claims for an impact at the Permian-Trias boundary marked by a geochemical anomaly in China have yet to be substantiated.

If we examine the developmental history of many animal groups, we find, again, crisis points disappearances and radiations, and extreme morphological punctuation, which follow characteristic patterns. Two examples are discussed. One, described by House, is from the ammonoids, an extinct subclass of cephalopods that underwent rapid development in the Devonian with two radiations, to be followed by further extinctions and major radiations into many complex morphological developments in the Carboniferous-Permian, the Trias, and the Cretaceous. At each crisis point all or nearly all specialized forms died out and a new radiation developed from smooth serpenticones that survived. All disappeared at the end of the Cretaceous, however, although their distant surviving relatives, the nautiloids, are rare examples of that particular experiment of a coiled shell cephalopod.

Kemp has described my other example, the evolutionary history of the mammal-like reptiles from the earlier Carboniferous to the end of the Trias. They underwent three radiations followed by extinction. In each they developed into large and small herbivores and large and small carnivores. At each crisis point, while some lineages may have continued for a time, only the small carnivores provide the ancestors of the next development, and again diverged into the same four general types. At the final extinction, at the end of the Trias, a small carnivore lineage developed into the mammals.[4]

The Driving Force of Evolution

It is clear that environmental change controls evolution at all levels, but when assessing the mechanisms of change, it appears that a difference in degree leads to a difference in kind. While Darwinian principles must still apply at any level, some of the gradualism of orthodox genetic theory must be sacrificed. Small evolutionary changes lead inevitably to specialization, as organisms adjust to a stable or slowly changing environment. Big changes in the environment, particularly if abrupt, bring about major transformations in life. Massive biomass extinctions provide large empty environments in which radiation and repopulation can take place rapidly. What is killed may be largely accidental, i.e., non-Darwinian, but what forms survived is important, and, in hindsight, they appear to have been ready for it. Repeatedly we see the unspecialized: the smooth serpenticone ammonoid, the small insectivore, forming the opportunistic root stock for vigorous divergence into the new environment. Not all specialized forms vanished simultaneously, but again, we find that the survivors continued in a restricted environment without major divergence into new lines, such as the trilobites from the latest Devonian to the Permian, or, today, the coelacanths and rhyncocephalians, both survivors from the Mesozoic.

To match this attack on gradualism we require the vigorous genetic mechanism provided by molecular biology: McClintock's power of the genome, Dover's molecular drive, and the idea of horizontal gene transfer add a new dimension to orthodox theory. Coupled with the empty environments provided by the physical punctuation we have been discussing, the term macroevolution takes on new meaning.

An Anthropocentric View

Now I turn again to man and try to put him in perspective. We need a distorted scale to chronicle our emergence, so here is an anthropocentric view of the universe (Fig. 5), plotted over eight orders of time. Notice that the crisis points accelerate, firstly in decreasing taxonomic rank to *Homo*, and then in behavioural crises. The agricultural revolution marks the end of man as an ecological animal; the Greeks suggested that perhaps there are laws of nature to discover; the

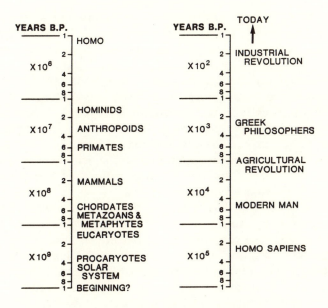

Figure 5 An anthropocentric view of the universe

industrial revolution, when man began to apply such ideas seriously, was followed quickly by the hygiene revolution, the population explosion, and the walk on the moon. The acceleration is interesting, and may represent another law of evolution. Very recently, Russell and, quite independently, Wyles and others have suggested that accelerated evolution in certain groups may be correlated with greater encephalization (relative brain size), leading to behavioural innovation and, by implication, intelligence. Encephalization has been shown to have operated among cephalopods, song birds, dolphins, primates, and other groups. Man would appear to be an end member of this process and to have further advanced and accelerated in the particular behavioural innovation that could be described as intellectual activity – which is why we are here today. It is this final innovation that enables us to contemplate the universe, and to realize with dismay, that we may be alone.[5]

Conclusion

Life appears to have originated very early in Earth's history, certainly by 3.5 Ga, probably some time before that, and possibly not long after the formation of the crust at 4.2 Ga. Since then, Gaia has preserved a stable environment for life within narrow limits, although several major forces, internal and external, have sporadically disturbed the environmental norms. Two complementary evolutionary mechanisms developed: slow evolutionary changes, in equilibrium with the environment, and large changes caused by major environmental disruptions. Evolution appears to be an intricate accelerating system of general processes with chance the driving force, but not the determinant. Over the long time life has existed on Earth, it is difficult to indicate any particular crisis point as being more improbable than any other. The timing of major environmental shocks may be largely stochastic, but within the overall time scale they are a normal component of the total environmental flux.

Is this appearance of design merely the result of looking back at entirely chance happenings, the effects of *ex post facto* observation? If so, we must embrace the alternative that out of a large (infinite?) number of universes, ours happens to be the one where all this took place. I promised to return to Gosse. Is postulating infinite universes, all governed by chance, any different from Gosse's universe with evidence created only in God's mind? Both suggestions are metaphysical, in that each contains a denial of the possiblity of their being tested. The terminology differs, but can it seriously be claimed that one is more acceptable than the other? I shall close, not with an answer, but with a plea that man should act, in the absence of any evidence to the contrary, as though he were unique, the sole observer of the universe. Such a moral imperative might finally provide us with the necessary stimulus to take steps to ensure our survival.

NOTES

1 P.C.W. Davies, *The Accidental Universe* (Cambridge: Cambridge University Press, 1982), contains references to Everett, Eddington, Pantin and Hoyle. I. Prigogine and I. Stengers, "A New Model of Time, a New View of Physics," in *Models of Reality: Shaping Thought and Action*, ed. J. Richardson (Maryland: Lomond Publications, 1984), contains useful references including an earlier paper by Prigogine. J.E. Lovelock's *Gaia: A New Look at Life on Earth* (Oxford: University Press, 1979) is important to the present discussion and has references to Lynn Margulis's work on cellular evolution, and others.

2 Schopf et al. refers to *Earth's Earliest Biosphere: Its Origins and Evolution*, ed. J. William Schopf (Princeton: Princeton University Press, 1983), which

contains many relevant articles, including those by Preston Cloud and Jan Veizer as well as many references. A. Seilacher's views on the Ediacara fauna are to be found in *Patterns of Change in Earth Evolution*, ed. H.D. Holland and A.F. Trendall, a Dahlem Workshop Report (Berlin: Springer-Verlag, 1984). This report also includes papers by E.M. Shoemaker and A.G. Fischer, and others of interest, as well as many bibliographical references. H.B. Whittington and his research team are close to completing a full revision of the Burgess Shale biota and their results are widely published; a summary with references may be found in *The Burgess Shale* (New Haven: Yale University Press, 1985).

3 Causes of extinctions, with references, are summarized in D.J. McLaren, "Bolides and Biostratigraphy," *Geological Society of America Bulletin 94* (1983): The impact theory is widely discussed in many articles in *Geological Implications of Impacts of Large Asteroids and Coments on the Earth*, ed. L.T. Silver and P.H. Schulz, Geological Society of American Special Paper 190, 1982.

4 The literature on episodic extinctions is diffuse and confused. The account given here is necessarily selective, but does not do justice to the complexity of the problems. Information may be found in some of the works cited above.

5 Because ideas on accelerating evolution are possibly controversial, full references are given to the papers mentioned: D.A. Russell, "Exponential Evolution: Implications for Intelligent Extraterrestrial Life," *Advances in Space Research* 9 (1983): 95–103. J.S. Wyles, J.G., Kunkel, and A.C. Wilson, "Birds, Behaviour, and Anatomical Evolution," *Proc. Natl. Acad. Sci. U.S.A.* 80 (1983): 4394–7.

RICHARD SWINBURNE

The Origin
of Consciousness

I begin by analysing my datum: I am concerned to explain the occurrence of conscious events. Events consist in the instantiation of properties in substances (that is, things) at times. My tie (substance) being red (property) now (time) is an event; so am I (substance) writing (property) today (time). (As I shall use the term, events include both continuing states and changes of state.)

Properties and events may be physical or mental. I understand by a physical property one such that no one subject is necessarily better placed to know that it is instantiated than is any other subject. Physical properties are public; there is no privileged access to them. Thus having a mass of ten pounds, being eight feet tall, and being square are all physical properties. So too are the typical properties of neurones in the brain – being in such-and-such an electrical state or releasing a transmitter substance. Anyone who chooses can find out as surely as anyone else whether something is eight feet tall, or in a certain electrical state. Physical events, as I shall use the term, are those that involve the instantiation of physical properties alone. Mental properties, as I shall understand the term, are ones to which one subject has privileged access, about which he or she is necessarily in a better position to know about than is anyone else. Properties such as being in pain or having a red after-image are mental, for any person in whom they are instantiated is necessarily better placed to know about them than is anyone else. For whatever ways you have of finding out whether I have a red after-image (by inspecting my brain state or studying my behaviour, for example) I can share; and yet I have an additional way of finding this out – by my own awareness of my own experience (an awareness that may or may not be fallible). Mental events are events that involve the instantiation of (one or more) mental properties (for instance, John being in pain at midday yesterday).

There are properties, and so events, that are mental on this definition, and that can be analysed in terms of a physical component and a mental component. These properties and events we may call mixed mental properties and events. Thus "saying 'hello'" is a mixed property, for my saying "hello" consists in my intentionally seeking to bring about the sound "hello" (a mental event) causing the occurrence of the sound "hello" (a physical event). It is the property of the instantiation of a certain mental property, causing the instantiation of a certain physical property, that is possessed by the substances that are people who say "hello." Those mental properties that cannot be analysed in part in terms of a physical component I term pure mental properties. "Being in pain" and "desiring to eat" are such pure mental properties. In talking henceforward of mental properties I shall understand thereby (unless I specify otherwise) pure mental properties; and I shall understand by mental events the instantiations of pure mental properties.

Among pure mental events are conscious events, events of which the subject is to some extent aware while he is having them. Sensations are an important class of conscious events. Sensations include the sensory content of perception – by experiencing patterns of colour in my visual field, hearing noises, feeling pressure (minus any beliefs about how things are in the outside world – that there is a table present, say – that are mediated by the patterns of sensation in perception). They include also experiences similar in content to those had in perception, such as those of dreams and hallucinations, and the pale imitations of perception in memory images and imagined images. They include also sensations of taste or smell and pains, and sensations of hot or cold in parts of the body.

Thoughts are another class of conscious events. By thoughts I mean those datable conscious occurrences of particular thoughts that can be expressed in the form of a proposition. Often these are thoughts that occur to a subject, flit through his mind, or strike him without the subject in any way actively conjuring them up. It may occur to me that today is Tuesday, or that I have a receptive audience, or that the weather is cold. A subject's awareness of his or her beliefs, and views of how the world is, come as thoughts. Perception involves, as well as sensations, the acquisition of belief (of which the subject is sometimes conscious) about how the world is surrounding the subject. (In so far as beliefs are states that continue while the subject is unaware of them, they are mental – because of the privileged access to them – but not conscious events.)

A further class of conscious events are what I shall call "purposings." By a subject's purposings I mean his endeavourings to bring about some events, meaning so to do. Every intentional action, everything that an agent does, meaning to do it, consists of the agent's bringing about some effect, or trying but failing to do so. Yet when the agent brings about some effect, the active contribution may be just the same as when, unexpectedly, he tries but fails to bring about the effect. When I move my hand, I bring about the motion of my hand. Normally this action involves no effort and is entirely within my control. But on some occasion, I may find myself unexpectedly paralysed. My active contribution is the same as when I move my hand successfully, yet because I fail, we say that what I did was to "try" to move my hand. Normally we speak of "trying" only when effort or failure is involved. Yet we do need a word that covers both the trying that involves effort or failure, and an agent's own intentional contribution to an action, when the performance is easy and successful. For this intentional contribution, for an agent's setting himself or herself to bring about some effect (even when effort or failure is not involved) I shall use the word "purposing."

A person's conscious life also includes from time to time awareness of desires, those in-built inclinations to do certain actions (or, in my terminology, to have certain purposings). (In so far as desires continue while the subject is unaware of them, they are like beliefs, mental states but not conscious events.)

All the conscious events that I have described, apart from sensations, are intrinsically propositional in character; that is, they consist in an attitude towards a state of affairs under a certain description. A thought is a thought that so-and-so is ("today is Thursday"); a purposing is a purposing that so-and-so occur; and so on. Sensations by contrast are unconceptualized experiences; though normally, in perception, they accompany the acquisition of beliefs about the external world. When I see a table, I have a coloured visual field (sensation) and I acquire thereby the belief that, causing it, is a table.

It seems to me evident that mental events, and especially conscious events, are distinct events from brain events or other physical events. To have a red sensation is not the same event as to have one's C-fibres fire or any other going-on in the brain. For each consists in the instantiation of a different property (redness; a certain distribution of electric charge) in the same subject at the same time. One event may cause the other, but there are two distinct goings-on. A Martian visitor who discovered everything physical that was happening in the human brain (in the way of redistribution of chemical matter and electric charge) would want to know whether or not humans felt anything when you kicked them and they screamed, whether they purposed their arms to move, whether they had thoughts about the world, and so on. A full history of a human being would list mental events as well as physical events. Much philosophical ink has been spent in trying to construct arguments to deny what seems to stare us in the face – that conscious events are distinct events from brain events.[1] Unfortunately, I do not have the space here to marshal counter-arguments, though I have done so elsewhere.[2] I shall therefore assume what I hope will be apparent – that consciousness is something distinct from what goes on in the brain, albeit closely connected causally to it. The conscious life, unlike the life of the brain, is rich in colour, feeling, and meaning.

That we ourselves are in the above respects conscious is immediately evident. It is not much less evident that other humans besides ourselves are conscious. It is fairly evident that the higher mammals are conscious. It is quite evident that inanimate matter is not conscious. At some stage in evolutionary history consciousness first appeared. I shall not speculate when that was. No doubt the various facets of consciousness arrived on the evolutionary scene at different moments of history.

Many of our sensations are evidently caused by brain events, and the latter are normally caused fairly directly by more remote bodily events (such as light rays impinging on eyes). Plausibly, too, many thoughts (especially thoughts that express the content of beliefs acquired in perception) are caused directly by brain events. We do, however, believe some of our thoughts to be caused by other of our thoughts (as when we do a piece of mental arithmetic). We also believe that our purposings are often efficacious – that my purposing to move my arm, say, causes my arm to move, and it can only do that by causing a brain event that causes the arm to move. The epiphenomenalist denies that there is any "downward" efficacy; he claims that when purposing appears to cause bodily events, what is really happening is that the two have a common cause, a brain event that causes a purposing to move an arm and, slightly later, the arm movement; it is this, he claims, that produces the illusion of purposive efficacy. It does not seem that way. And no one can be a consistent epiphenomenalist and live. For to purpose to do something is to do what one believes to be likely to bring about the event purposed. To try to move one's hand is to do what one believes is likely to be causally efficacious in bringing about the hand motion. An epiphenomenalist, however, believes that nothing mental has effect in the physical world, and hence cannot perform any mental act that is believed to have such an effect. So the epiphenomenalist cannot try to move a hand, or do any other intentional public action; the consistent epiphenomenalist must cease to try. But no one will do that, and that is because everyone really believes that their purposes are efficacious. I appeal to consensus in making the assumption that purposings affect brain states.

So we get the following picture of mind-body interaction (in the sense of the interaction of the conscious events and brain events of an agent; I am not assuming here that minds are substances): brain states cause many sensations and some thoughts; many purposings cause brain events; and perhaps some conscious events cause (in part) other conscious events. Can we have a scientific explanation of how the evolution of physical systems gave rise to conscious events that began to interact with physical events and direct the development of those systems?

II

The first thing we would need is an explanation of how in the present in humans and many other animals certain brain states give rise to certain conscious states, and conversely. For this purpose I shall make the normal (though very ill-evidenced) simplifying assumption of one-many simultaneous mind-brain correlation. That is, for each kind of conscious event there are one or more kinds of brain event, such that whenever one of the latter occurs, the former occurs simultaneously and conversely. Thus, given a specific kind of sensation, a round red image of a certain hue and size against a black background, say, there will be one or more kinds of brain events such that whenever one of the latter occurs, the image occurs; but the image never occurs without one of the latter occurring. In principle it will then be possible for scientists to compile a very, very long list of the correlations between brain events and conscious events, stating which conscious event occurs (a blue elliptical image, for another example) when a given brain event occurs. The correlations would be generalizations, stating that all brain events of this kind occur at the same time as conscious events of that kind (that is, correlations of the form "always whenever this brain event, that conscious event"). The correlations would be established by noting which brain states occurred when observers reported having some conscious event and not otherwise. To get exceptionless lists scientists would probably need to suppose that on some few occasions, to mislead the scientist or through inadequate grasp of sensory vocabulary or through misobservation, observers misreported their sensations.

But why do certain brain events give rise to conscious events and others do not, and why do brain events give rise to the conscious events that they do rather than to other events? Why does a certain brain event give rise to a pink image rather than to a smell of roses or to a thought that $2 + 2 = 4$? Physics and chemistry cannot possibly explain these things, since pink looks, rosy smells, and mathematical thoughts are not the sort of thing physics and chemistry deal in. These sciences deal in the physical (public) properties of small physical objects, and of the large physical objects that they come to form – in mass and charge, volume, and spin.

But could not physics and chemistry be enlarged so as to become a super-science dealing with both physical and mental properties, and providing explanations of their interactions? I do not think so, for the following reason. To give a scientific explanation of why some event occurred you have to show that, given the previous state of affairs, the laws of nature are such that it had to occur. You explain why the planet today is in this position by stating where it and other heavenly bodies were last month (the initial conditions) and how it follows from Newton's laws of motion and gravitational attraction that those initial conditions would be followed a month later by the planet being where it is today. To provide a scientific explanation you need laws of nature. Laws state what must happen of natural necessity to all objects with certain properties; that is, they claim, that some generalization "all As are B" (for example, As have certain properties or do so-and-so) holds of natural necessity. And what will be the evidence that some such generalization does hold of natural necessity? First, of course, no As must have been observed to be not-B; and secondly, normally, many As must have been observed to be B. But that is not enough. All ravens so far observed have been observed to be black. But that is quite inadequate evidence for supposing that "all ravens are black" is a law of nature. We need evidence to suppose that all ravens so far observed being black is no mere accident of local conditions (caused by the chance that there has not actually occurred any mutation to produce a gene for whiteness in the genotype of raven) but that there could not be a black raven (however molecules were arranged, they could not give rise to a black raven). To show this of some generalization "all *As* are *B"* we need to show not merely that the generalization holds universally, but that it fits neatly into a scientific theory that is a simple theory with few simple purported laws, able to predict a vast range of phenomena. The grounds that people had in the eighteenth century for believing Newton's "law" of gravity – that all bodies attract each other in pairs with forces proportional to the product of the masses of each and inversely proportional to the square of their distances apart – to be indeed a law of nature, was that it was a very simple law that fitted into a theory with four "laws" in all, the others being equally simple, that together were able to predict a vast range of phenomena. The law of gravity ($F = mm^1/r^2$) is simple because the distance is not raised to a complicated power (e.g., we do not have $r^{2.003}$), there is only one term (e.g., we do not have $mm^1/r^2 + mm^1/r^4 + mm^1/r^6$), and so on. The law is also simple for the reason that, given this law and only given this law, the total force exerted by a body on a hollow sphere of given uniform thickness and density centred on the body remains the same, whatever the inner radius of the sphere. The theory, with four such simple laws, was able to predict with great accuracy the behaviour of bodies of very different kinds in very different circumstances – the motions of planets, the rise and fall of tides, the interactions of colliding bodies, the movements of pendula, etc. Einstein's laws of General Relativity are, of course, more complicated than Newton's, but Einstein claimed for them that they had great coherence and great mathematical simplicity and elegance, at any rate in comparison with the complexity of the data they were able to explain. I conclude that the evidence that some generalization is a law, as opposed to a mere generalization (holding only perhaps in a limited spatio-temporal realm, or by accident or coincidence), is that it fits into a simple coherent theory that predicts successfully a vast range of diverse data.

Now a scientist, I have assumed, could compile a long list of correlations between brain events and conscious events. But to explain those correlations we need by our principles to establish a much smaller set of purported laws, from which it follows that this kind of brain event has to be correlated with a red sensation, that one with a blue sensation, this one with a purposing to sing a high note, and that one with a purposing to sing a low note. The purported laws would need to fit together into a theory from which we could derive new correlations (predict, for instance, some totally new sensation to which some hitherto unexemplified brain state would give rise). If our purported laws could do all that, that would be grounds for believing them to be indeed laws, and so for believing that we had got a scientific explanation of the occurrence of sensations. Then we would be able to say why sodium chloride tastes salty rather than sweet in terms of the brain event that tasting sodium chloride normally causes, having a natural connection (stated by our theory) with a salty tang. Mere correlation does not explain, and because it does not explain, you never know when your correlations will cease to hold. Because you have no explanation of why all ravens are black, you may reasonably suspect that tomorrow someone will find a white raven.

The list of correlations is like a list of sentences of a foreign language that, under certain circumstances, translate sentences of English, without any grammar or word dictionary to explain why those sentences are under those conditions correct translations. In the absence of a grammar and dictionary you do not know when those translations will cease to be accurate (maybe "blah blah" only refers to the Sovereign when the Sovereign is male). Nor can you translate any sentence other than the ones listed.

But why should not the scientist devise a theory showing the kinds of correlation discussed to be natural ones? Why should the scientist not postulate entities and properties from whose interactions, the laws of which are simple, it would follow that you get the correlations that you do between brain events and conscious events? Although it is theoretically possible that a scientific theory of this kind should be created, the creation of such a theory does not look a very likely prospect. Brain events are such different things qualitatively from pains, smells, and tastes, thoughts and purposings, that a natural connection between them seems almost impossible. For how could brain states vary except in their chemical composition and the speed and direction of their electro-chemical interactions, and how could there be a natural connection between variations in these respects and variations in the kind of respects in which tastes differ – say the difference between a taste of pineapple, a taste of roast beef, and a taste of chocolate – as well as the respects in which tastes differ from smells and smells from visual sensations, and sensations from thoughts and purposings. There does not seem the beginning of a prospect of a simple scientific theory of this kind and so of having established laws of mind-body interaction as opposed to lots of diverse correlations, which, just because they are unconnected in an overall theory, are for that reason not necessarily of universal application. If we cannot have scientific laws we cannot have scientific explanation. The scientist's task of giving a full explanation of the occurrence of a subject's conscious events seems doomed to failure. For a scientific theory with detailed laws could not be simple enough for us to have reasonable confidence in its truth (and so its universal applicability).

The history of science is, it is true, punctuated with many great "reductions," of one whole branch of science to another apparently totally different, or "integration" of apparently very disparate sciences into a super-science. Thermodynamics, dealing with heat, was reduced to statistical mechanics; the temperature of a gas proved to be the mean kinetic energy of its molecules. Optics was reduced to electro-magnetism; light proved to be an electro-magnetic wave. And the separate sciences of electricity and magnetism came together to form a super-science of electro-magnetism. How is it that such integrations can be achieved if my argument that there cannot be a super-science that explains both mental events and brain events is correct?

There is a crucial difference between the two cases. All other integrations into a super-science, of sciences dealing with entities and properties apparently qualitatively very distinct, were achieved by saying that really some of those entities and properties were not as they appeared to be, by making a distinction between the underlying (not immediately observable) entities and properties and the sensory properties to which they gave rise. Thermodynamics was originally concerned with the laws of temperature exchange, and temperature was supposed to be a property inherent in an object, that you felt when you touched the object. The felt hotness of a hot body is indeed qualitatively distinct from particle velocities and collisions. The reduction was achieved by distinguishing between the underlying cause of the hotness (the motion of molecules) and the sensations that the motion of molecules cause in observers. The former falls naturally within the scope of statistical mechanics – for molecules are particles; the entities and properties are not of distinct kinds. But this reduction has been achieved at the price of separating off the sensory from its causes, and only explaining the latter. All "reduction" of one science to another dealing with apparently very disparate properties has been achieved by this device of denying that the apparent properties (such as the "secondary qualities" of colour, heat, sound, taste) with which one science dealt belonged to the physical world at all. It siphoned them off to the world of the mental. But then when you come to face the problem of the sensations themselves, you cannot do this. If you are to explain the sensations themselves, you cannot distinguish between them and their underlying causes and only explain the latter. In fact the enormous success of science in producing an integrated physicochemistry has been achieved at the expense of separating off from the physical world colours, smells, and tastes, and regarding them as purely private sensory phenomena. The very success of science in achieving its vast integrations in physics and chemistry is the very thing that has made apparently impossible any final success in integrating the world of the mind and the world of physics.

There is little prospect for a scientific theory of the origin of sensations. And what goes for sensations goes for conscious events of other types. There is not the ghost of a natural connection between this brain event and that thought, so that we could understand how a change of this electro-chemical kind would give rise to a thought that $2 + 2 = 5$ as opposed to a thought that $2 + 2 = 4$, the thought that there is a table here rather than the thought that there is no solid object here. There is a vast qualitative difference between thoughts with their in-built meanings (their intrinsic propositional content) and mere electro-chemical events, so that it seems impossible to construct a theory in terms of which the various correlations between brain states and thoughts follow naturally, that was simple enough for us to expect to make successful predictions about which new thoughts would be correlated with hitherto unexemplified brain states. Only if this were done would the theory be one that we would be justified in believing to provide true explanations of the occurrence of thoughts. And what goes for thoughts goes also for purposes and other conscious events with intrinsic propositional content. There is no prospect of a scientific theory of why a purpose to move my hand should cause this neurone to fire rather than that one.

III

Let us, however, take it as given that by some means or other it comes about that brain events of certain kinds give rise to conscious events of certain kinds and, conversely, that the long list of correlations is operative. There are two remaining issues that science may have some prospect of explaining. First, how did it come about that there were organisms with brains with a repertoire of states of a kind to cause conscious events and be caused by them?

It is possible that there is a scientific explanation of this fact, though it is not easy to see how it would run. An orthodox Darwinian account would claim that genetic mutations or rearrangements caused the existence of genes that caused the growth of brains capable of mental interactions of certain kinds; organisms with such brains survived either because there was a selective advantage in having a conscious life or because there was a selective advantage in other characteristics also brought about by the same genetic material as gave rise to a conscious life. It is not easy to see what the selective advantage of having a mental life is. Just consider sensations. This system of ours in which sensations are causally intermediate between stimulus and response seems to have no evolutionary advantage over a mechanism that produces the same behavioural modifications without going through sensations to produce them. It looks as if there can be mechanisms of the latter kind – are not the light-sensitive, air-vibration-sensitive machines that are beginning to be made just such machines? Maybe there is a discoverable physico-chemical explanation of why there cannot be (or is not very likely to be produced) a set of genes that will give rise to such a machine; or perhaps an organism with sensations has a less cumbersome process for modifying its behaviour (in response to bodily damage, say) than such an unconscious machine, and for that reason has an evolutionary advantage. The former, if true, would be far easier to show than the latter. However, along one of these lines there are possibilities for a Darwinian account of why organisms with a capacity to have sensations have an advantage in the struggle for survival.

The same point applies to conscious events of other kinds. There is no apparent advantage possessed by an organism who has a belief articulated in conscious thought that there is a table here, as opposed to a mere disposition to avoid bumping into the table and to put things on it (such as presumably is possessed by robots). What advantage is there in the mental awareness as opposed to the unconscious disposition? As with sensations, we can sketch a possible answer – perhaps organisms with conscious beliefs and so on will have less cumbersome processes for reacting to their environment than organisms without a mental life who react in the same way, and for that reason the former have greater survival value. (Robots perhaps need more bits and take up more space.) Or maybe it is not possible to have a set of genes produced by recombination of DNA molecules (as opposed to a silicon chip) that will give rise to an organism with complicated abilities to react to the environment, without that organism having beliefs about it and other mental attitudes towards it.

There are prospects for a Darwinian explanation of how animals have evolved with brains of the sort to give rise (by an inexplicable process) to consciousness, though the prospects are not very bright ones. In the absence of a Darwinian explanation, there might be a fairly ordinary scientific explanation of some other kind. Maybe DNA has got an in-built propensity to give rise to genes of the requisite type during the course of millions of years of recombination; maybe there is an orthogentic tendency to evolve in this direction.

IV

We have noted, then, one crucial all-important question that is utterly beyond the powers of Darwinism or apparently science itself to answer – why do certain brain events give rise to certain mental events, and conversely? – and one question on which there are some possibilities for a Darwinian or other scientific answer. There is a third question, to which Darwinism can provide a clear and obviously correct answer as regards conscious beliefs, purposes, and desires. This is the question: given the existence of mind-brain correlations, and given that organisms with a conscious life will be favoured in the struggle for survival, why are the brain events that cause and are caused by conscious events connected with other bodily events and extra-bodily events in the way in which they are? Take beliefs (of which the subject is aware in consciousness). Why is the brain connected via the optic nerve to the eye in such a way that the brain event that gives rise to the belief that there is a table present is normally caused to occur when and only when there is a table present? The answer is evident: animals with beliefs are more likely to survive if their beliefs are largely true. False beliefs, such as those about the location of food or predators, will lead to rapid elimination in the struggle for survival. If you believe that there is no table present, when there is one, you will fall over it, and so on. Those in whom the brain events that give rise to beliefs are connected by causal chains to the states of affairs believed are much better adapted for survival than those whose belief brain states are not so connected and who in consequence tend to have false beliefs. Many animals have a built-in mechanism for correcting in the light of experience any tendency to acquire false beliefs by a certain route – finding frequently that an object of a certain kind that looks like food is really not food, for example, they cease to acquire the belief that there is food in front of them when they receive visual stimuli from an object of that kind. Such animals are more likely to survive and produce offspring. A similar account can be given of why the brain events produced by purposes give rise to the movements of body purposed. If, when I tried to move my foot, my hand moved instead, predators would soon overtake me. Similarly, given that I am going to have desires caused by brain events, there are evolutionary advantages in my having some under some circumstances rather than others under other circumstances – desire for food when I am hungry rather than when I am satiated, say. I do, however, have some doubt as to whether there can be a satisfactory explanation of why the brain events that give rise to sensations (as opposed to thoughts and purposes) are connected to bodily and extra-bodily events in the way that they are. What is the selective advantage in ripe tomatoes and other objects normally called "red" giving rise to red sensations, and unripe tomatoes and other objects normally called "green" giving rise to green sensations rather than conversely?

In summary, then, the evolution of consciousness in animals is a matter of: (1) there existing certain physical/mental correlations (brain events of certain kinds being correlated with conscious events of certain kinds); (2) there having evolved animals with brains, having a repertoire of events of the kinds correlated with conscious events; and (3) those animals having their brains "wired into" their bodies so that peripheral and extra-bodily events are causally connected with consciousness in certain familiar ways. Darwinian or other scientific mechanisms can explain quite a lot of (3), and possibly some of (2), but neither Darwinism nor any other science has much prospect of explaining (1). Hence to explain adequately the origin of the most novel and striking features of animals (their conscious life of feeling, choice, and reason) probably lies utterly beyond the range of science.

I conclude that the process of animal evolution, apparently so regular and predictable, is yet in the respect of those all-important properties of animals (their mental life that makes them, like humans, deserving of kindness and reverence, and that makes them also interact with ourselves) not fully explicable scientifically, and not likely ever to be.

V

I have argued that science cannot fully explain the evolution of a mental life. This is because, so far as we can see, there is no law of nature stating that physical events of certain kinds will give rise to correlated mental events and, conversely, there is nothing in the nature of certain physical events or of mental events to give rise to the correlations. Yet there are so many evident regular correlations between consciousness and the brain that their orderliness cries out for explanation. Is there any other way of explaining them?

There are two basic and very different ways of explaining phenomena. There is the scientific way of explaining an event by a preceding
initial set-up and laws of nature, such that the description of the latter
two entails the description of the event to be explained; laws
themselves are explained by being deducible from higher-level laws
(and perhaps also some general description of a pervasive feature of
the Universe on the small or large scale). By contrast there is personal
explanation, whereby we explain an event as brought about intentionally by some agent, for its own sake or for the sake of some further
purpose. Much human behaviour is explained in the latter way, and
such "explanation" is recognized as explanation, as showing what it
was that brought about the event in question.[3] But personal explanation shows an event occurring because an agent sought to bring about a
goal, rather than because prior blind forces brought it about.

A possible personal explanation in this case is that God, the power
behind nature, brings it about that the brains of men and some animals
are connected with a conscious life in regular and predictable ways.
God, an omnipotent, omniscient, perfectly free, and perfectly good
source of all, would need to be postulated as an explanation of many
diverse phenomena (including the operation of laws of nature themselves) to make his existence probable. But the ability of God's actions
to explain the otherwise mysterious mind-body connection is just one
more reason for postulating God's existence.

The suggestion is that God has given to each human brain and some
animal brains a limited nature, as it were; a limited nature such that in
the circumstances of normal embodiment it keeps a conscious life
functioning and interacting with the brain in predictable ways without
the brain having that nature deriving from any general law of
brain/mind connection, and determining in general under what
circumstances there occur conscious events of different kinds.[4] God,
being omnipotent, would have the power to give to the brain such a
limited nature, to produce intentionally those connections which, we
have seen, have no natural connection. And God would have a reason
for so doing: to give to human beings beliefs, thoughts, desires, and
sensations caused in regular ways by brain states, and purposes that
cause brain states in regular ways, would allow them to acquire
knowledge of the world and to make a difference to it by conscious
choice – to allow them to share in the creative work of God himself. That
there are animals who also acquire some true beliefs and make some
choices, and interact with each other and with men in so doing, is also a
good thing that an omnipotent God would have reason for bringing
about.

A God would have the ability and a reason for bringing about such connections. The occurrence of the conscious life and its mode of functioning (under the limited conditions of embodiment in bodies with brains), which otherwise are likely to remain totally mysterious, can be explained in terms of divine action.

NOTES

This paper is based on material contained in my book, *The Evolution of the Soul* (Oxford: Clarendon Press, 1986), particularly ch. 10.

1 See, for example, most of the papers in *The Mind-Brain Identity Theory*, ed. C.V. Borst (London: Macmillan, 1970); D.M. Armstrong, *A Materialist Theory of Mind* (London: Routledge and Kegan Paul, 1968); H. Putnam, "Minds and Machines" and "The Mental Life of Some Machines," in his *Philosophical Papers*, Vol. 2: *Mind, Language and Reality* (Cambridge: Cambridge University Press, 1975); and D. Davidson, "Mental Events," in his *Essays on Actions and Events* (Oxford: Clarendon Press, 1980), 207–25.

2 *The Evolution of the Soul*, ch. 3; and "Are Mental Events Identical with Brain Events," *American Philosophical Quarterly* 19 (1982): 173–81.

3 Personal explanation is entirely different in pattern and cannot be reduced to scientific explanation. On this see *The Evolution of the Soul*, ch. 5; or my *The Existence of God* (Oxford: Clarendon Press, 1979), ch. 2. The next few paragraphs summarize the argument for the existence of God from consciousness given in *The Existence of God*, ch. 9.

4 My account leaves it open whether God confers this limited nature on evolved animal brains, once they have them; or, as it were, determines in advance that brains with a certain repertoire of states will be the ones that give rise to consciousness, leaving it to Darwinian or other scientific processes to bring it about that animals and humans have brains of that kind.

DOREEN KIMURA

The Origin of Human Communication

On learning the topic of this volume, a philosopher I know said: "Surely if someone was going to design people, he could have done a better job?" Many people, however, who are quite willing to admit that a daisy or even a monkey could have come about by chance, believe that human beings must have come about by design. Scientists are apparently not exempt from this attitude, and non-biological scientists, in particular, often conjecture that the complexity of human thought and ability precludes their emergence through natural selection alone. Certainly it sometimes seems that the pervasiveness of human communication, for example, defies rational explanation.

The origins of primates, however, and especially of man and the great apes, surely reveal a basis for the gradual selection of basic abilities related to calculating the distance of a moving object; accurately communicating a source of food; discriminating a stranger from a group member; avoiding a predator; and responding to signs of illness or hunger in an infant. Such abilities would, in the long ago world in which our forebears lived, have tended toward survival and towards reproductive success. This is elementary evolutionary theory, and it lends itself to explaining more elaborate behavioural sequences, such as finding a moving prey, launching a missile at it, and hitting it, or communicating exactly the kind and location of a supply of food. These abilities are impressive but surely not inexplicable. The by-products of these abilities should not then be surprising either – mathematical and spatial ability, say, and communicative activity that goes far beyond what appears to be merely *needed*.

It is also possible that as mobile organisms with some control over the environment, we often would have been in situations where problem-solving abilities would have proved advantageous. Some of the puzzles posed might have been: how does one get across the fast-moving river? where does the ice come from? what makes the birds migrate? and so on. Being able to answer these questions would have all made for survival, directly or indirectly, and so the abilities would have been selected for in the process of evolution. It seems likely that early man may have also sought to discover his own origins, for explanations and beliefs are the natural consequence of dealing with the environment in a problem-solving way.

The recent history of our own surmises about the nature of man and the universe, in the western world at least, has been characerized by a steady replacement of religious, or non-material explanations, by material ones. This trend seems likely to continue, as long as we are able to maintain and transmit our knowledge to other con-specifics. But the "accuracy" of our beliefs need not have any consequence for the state of man on earth, unless such explanations result in interaction with the environment. Natural selection cannot operate on mental events, unless there is some overt consequence. The more removed earlier explanations were from any consequence in the real world, the less likely was the tendency to engage in them to be selected out or selected in. Thus, unless supernatural explanations significantly affected early man's adaptation to the environment, they could not be a disadvantage; or, at least, any disadvantage would not outweigh the advantage of possessing an explanation-seeking intelligence. The veracity of such beliefs is, in itself, irrelevant to survival and to reproductive success. The idea that there must be some survival function to every aspect of behaviour is mistaken. Even the papers in this volume may be a superfluous intellectual manifestation of some quite important abilities.

There remains, however, some room to share Dr Swinburne's puzzlement over the phenomena of mental events or consciousness, within a scientific framework. I refer below to what may be a plausible scientific explanation for the occurrence of states that we call "mental events" or "consciousness," which emphasizes their importance for social interaction. But because little or no research has been done on the subject of consciousness, we must be satisfied with data that have a less direct bearing on it. I want to suggest that studying the level of language capacity may be a rough guide to the level of consciousness, both within and across species. And of course studying the evolution of language will in itself tell us something about the nature of human intellect.

Consider that for many purposes the term "consciousness" can be translated as "capable of being labelled." If, for example, on interrogating someone about an event, one asks, "What were you conscious of at the time?" the question and the answer both imply that conscious events can be put into words. If, on the other hand, words cannot be used adequately to describe an event, for example, how one learns to keep one's balance in riding a bicycle, we talk of these as "unconscious events." The only meaning of consciousness that may escape this translation into the verbal is the immediate awareness of sensation or feelings, such as heat, pain, redness, music, and the like, which occupy one's thoughts at the moment without being labellable. Reflective consciousness, in the sense of thinking about one's sensations and feelings, and attributing motives or reasons, must require at least a rudimentary language, insofar as it involves symbolic function. The use of a label or designate (or some other kind of brain activity) to stand for a sensation or feeling presupposes symbolic function of a sort. Thus, to the degree that the sign-referent relationship is a basic characteristic of language, it seems reasonable to assume that what we mean by consciousness has developed in rough parallel to the development of language, or of language capability (Skinner 1976).

Many behavioural scientists have not in the past felt it necessary to invoke states such as awareness or consciousness, their reluctance to deal with this rather loose class of concepts being based on the fact that it did not appear to have any explanatory value for behaviour. Whether a person is conscious or not does not appear to make much difference in how we might account for his behaviour scientifically; in other words, there appears to be no need to provide materialist explanations for mental events separate from their behavioural manifestations.

However, the fact that human beings have the strong feeling that they possess a characteristic called "consciousness," and the fact that most behavioural scientists, if pressed, would be able to scale animal species along a consciousness-dimension, suggests that there may be a *behavioural* phenomenon here that is no less valid a field for scientific study than are other more obvious behaviours. This question was certainly addressed by radical behaviourism (Skinner 1976), and it has been reopened recently by Nicholas Humphrey (1984), who asserts that there *is* a function of reflective consciousness that is behaviourally manifest. He suggests that the ability to examine our motives allows us to build a model of behaviour that we can then apply to others; that we use the models we have created of our own behaviours to understand the behaviour of others. Having such a model available, however imperfect, has, according to this view, significantly greater predictive power for other people's behaviour than one could otherwise achieve. This ability could of course be of enormous survival value in a milieu where the social group was becoming significantly larger or more important.

The complex system of human communication is a behavioural system used primarily to inform and persuade. Communication provides a good model of the evolution of other human abilities, and it has been well researched. I shall be concerned primarily with language systems, though communication itself is a much broader term. But what I have to suggest about the origin of language systems is of a general enough nature that it could apply to non-verbal communication as well. I shall try to give an account of the evolution of human communication which makes it plausible that our current systems of communication have been present in basic form for a very long time – hundreds of thousands, if not millions of years; and that refinements toward our present systems of communication probably came about slowly by an incremental, though not necessarily even, series of changes.

Such an explanatory program involves significant assumptions about continuity. First, it implies a continuity between human and animal mentalities. Second, it implies the continuity that exists between the intellect of modern man and that of the oldest known hominids. In other words, the behaviours we emit during communication were probably based on a behavioural repertoire already in place, at least in rudimentary form, in hominids two or three million years ago. Such "precursor" behaviours were probably developed for some other function and were modified for use in communicating. I shall examine first the continuity with other animals and then the continuity with our hominid ancestors.

Continuity with Other Animals

To deal adequately with the similarities between ourselves and the great apes would require a review far beyond the scope of this paper; yet the assumption of a commonality is central to the argument that the behaviours we emit were selected from an array of pre-existing behaviours, because they were adaptive to the environment of our early forebears. Some of these behaviours should be shared by other primates, and the search for such parallels has been most intense in the chimpanzee and gorilla.

The tortuous and contentious history of the investigation of the chimpanzee's intellectual capacity might alone convince one of Humphrey's (1984) position that we understand very poorly any but human minds, because we have no internal models for any other. Our inadequate understanding of the human intellect should perhaps give us even more caution. The earliest pioneering studies on chimpanzees by Wolfgang Köhler (1959), done at the beginning of the century, posed the question of parallels in an essentially atheoretical way. He asked whether the chimpanzees he was studying solved problems and emitted behaviours in such a way that, if they were emitted by people, we would call insightful or reasoned. This apparently naive point of view actually is not without a certain sophistication; it precludes establishing criteria for the occurrence of a characteristic in a chimp that does not just as stringently apply to a human. The assumption that we know how human beings think is not correct in all instances.

The many pitfalls of working with non-human animals are related to the fact that human investigators do not share with non-human animals an identical biology, a common history, or a common means of communication. Imagine trying to find out how another human being, who has been reared by other animals (the "feral" person), thinks. The difficulties would be enormous, but they would still be much less than trying to discover this for a member of another species. Consequently, students of chimpanzee behaviour have had to be very inventive in order to find out how the chimpanzee views his world (usually, in fact, *her* world, because females have been more popular laboratory animals). While there is a great deal of discussion and argument about what intellectual capacities the great apes in fact have, I think the bulk of the research would convince the unbiassed reader that they do indeed possess abilities very similar in kind to our own, and of course others different from ours.

I shall survey some of these studies somewhat superficially, giving a more detailed account of those concerned with language abilities. Let us consider first the relatively simple question of personal identity, an appreciation of "self" and "other." A rather ingenious series of studies by Gallup (1977), employing mirrors, suggests that chimpanzees, with very little exposure to mirrors, identify themselves, and that they detect changes in their appearance induced by the experimenter. This contrasts with a failure to demonstrate this in baboons and in monkeys. It is also clear, I think, both from naturalistic observation and from rigorous laboratory studies, that chimpanzees can really co-operate to achieve a common aim – that is, they can vary their joint interactive behaviours in such a way as to achieve a remote goal (Goodall 1971; Savage-Rumbaugh, Rumbaugh, and Boysen 1978). Recently, some of the more innovative researchers in the field have even asked: "Does the chimpanzee have a theory of mind?" (Premack and Woodruff 1978). They have tried to discover if a chimpanzee can model internally another's state of mind when in a particular situation (represented by means of a videotape) and come up with an appropriate response. If they can, Premack and Woodruff argue, we may attribute to the chimpanzee a concept of "mind much like our own when in a similar situation." It is arguable whether or not all these attributes have been successfully demonstrated in chimpanzees, but there is no doubt that some of these behaviours do not occur in other primates such as monkeys, and therefore cannot be dismissed as merely experimental artifacts. They show a genuine between-species difference.

The most extensive and definitive work has been done in the area of communication and language. There are certainly both methodological and conceptual criticisms that can be levelled at much of the research, but I think the question whether the chimpanzee can communicate in rough parallel to human systems of communication must be answered in the positive. What does this mean? To employ a definition of Bennett's (1976): the best inference from work on language with the great apes suggests that they can intentionally inform and persuade other chimps. Thus they emit communicative behaviours only or primarily when a receiving chimp (or human) is known to be present. The appropriate behaviours are effective in conveying something that would otherwise not be conveyed to the receiver; this is indicated by the receiver's action or reaction. In other words, chimpanzees do not merely go through stereotyped movements, regardless of the presence or receptive behaviour of another.

The question whether the chimpanzee possesses the capacity for language depends of course on how one defines that term. Volumes have been written on the subject (see, for example, *Behavioral and Brain Sciences* 4 [1978], and deLuce and Wilder 1983). It is obvious, and perhaps trite, that one can define language in such a way that only human beings are capable of it. This attitude is possible because we are comparing quite different primate species and there is no question that they will differ in many ways. The strategy of the extreme humanist approach, when faced with the sometimes amazing feats of chimpanzees in problem-solving and/or communicative situations, is to respond, "Yes, perhaps they can do this, but not that, which only humans can do." Let us avoid this issue by admitting that the complexity, extent, and variety of human communication is probably unmatched by any other animal, and adhere to the question, what characteristics of our language system are nevertheless present in the great apes, and are these reasonably to be considered as precursor behaviours?

A list of the characteristics of communication systems compiled by Hockett (1960) still serves as a useful framework for discussion of language systems. Human language is said to have a number of characteristics which are only partly, or not at all, shared by non-human communication systems. Some of the more significant are: semanticity – the ability to use a sign or symbol to stand for a referent; arbitrariness – the sign does not necessarily resemble its referent; displacement – the ability to refer to things or events not present; productivity – the capacity to generate new, not previously used signs for statements; traditional transmission – learned from conspecifics, a characteristic that follows from arbitrariness; duality of patterning – a particular morpheme (or equivalent) will mean different things in different combinations (the "ba" in "bad" and "bat," for example). It is possible to add syntax – the order of signs in an utterance can change the meaning of the utterance (e.g., "Soon you will go"; "Will you go soon"). We can use these criteria as an outline for studying language in the great apes.

Since human communication uses the vocal musculature, it was natural that scientists interested in studying the language of non-human primates focused on their ability to use "speech." This excessively anthropomorphic approach was sometimes carried to an extreme when a chimpanzee was reared in a family with a human child, and many, as it turns out, fruitless hours were spent attempting to teach it human speech sounds. The achievement of chimpanzee Viki, who learned to utter about four intelligible words, though she apparently understood about fifty others, was the pinnacle of success (Hayes 1951). For several years of intensive training this is a poor showing indeed, and it seemed to indicate that the chimpanzee was hopelessly incapable of learning a language. However, these findings may have only shown that the chimpanzee's vocal tract is so different from ours that it is simply not capable of forming human speech sounds (Lieberman 1975). Hindsight might suggest that the investigators ought to have concentrated on getting the chimp to employ its own natural sounds for complex communication, but such an attempt has not been been seriously made since Yerkes tried to condition chimpanzee vocalizations in the 1920s. Moreover, the evidence that vocalizations in monkeys are subcortically mediated and apparently cannot be conditioned readily (Jürgens 1979) has provided little encouragement to this avenue of research.

Table 1 Human Speech in the Great Apes

	Training	Results
Witmer 1909	yes	"mama" (unvoiced vowel)
*Furness 1916	yes	"papa" "cup"
Kellogg 1932	no	–
Kohts 1935	no	–
Hayes 1951	yes (6½ years)	"mama" "papa" "cup" "up"? (single hoarse vowel for all)

*an orangutan; all others are chimpanzees.
(after Kellogg 1968)

There is, however, another major avenue of communication which human beings can and do employ under special circumstances, and that is a manual system of communication. When we vocalize, we tend to wave our hands and arms around in apparently aimless ways, as well as using them to describe, outline, or point in an informative way while talking. Such movements can be formed into a very complex and subtle language system in people who do not speak because they are deaf. Manual signing systems have survived among the deaf despite consistent suppression of those systems by hearing educators. The most studied of these sign languages is American Sign Language (ASL or Ameslan).

Ameslan is the fourth language most often used in the United States, ranking after English, Spanish, and Italian (Fant 1977). Its status as a "real" language, with the characteristics of a signal-meaning correspondence (Chomsky 1967), and the characteristics of language that Hockett (1960) has discussed, including recombinance and arbitrariness, has been thoroughly established (Klima and Bellugi 1979). In ASL the prevalent form of signing (the manual component of the transmitted signal) is a series of movements involving the whole arm, on one or both sides, in which a particular hand shape is moved through a pattern in a location specified with respect to the body (Stokoe, Casterline, and Croneberg 1965). For example, "pretty" is represented by one hand making a circular movement around the face (in clockwise direction as seen by the viewer), with the hand beginning the movement with all fingertips touching, then opening wide as it goes around the face, and ending with fingertips in contact with each other again. Roughly speaking, each sign corresponds to a concept, thing, or event, but there is no necessary equivalence with English words or indeed with words of any spoken language. (It is important to distinguish a natural sign from finger spelling. The latter is merely a direct letter-by-letter transcription from a spoken – and hence spelled – language into hand postures. Thus the word "in" is finger-spelled I, N, but the *sign* for "in" is quite different.)

Space does not permit a lengthy description of the richness of this language, but one example will do. There is a variation on an adjectival form that permits a distinction between an adjective, e.g. "wrong," and a related adjective showing a chronic tendency, such as "prone to be wrong." There are many other such variations, especially on verb forms. For a fuller exposition of these and other details of Ameslan, I refer the reader to Klima and Bellugi (1979).

Language in the Great Apes

Kellogg (1968), one of those psychologists who adopted a chimpanzee and attempted to teach it speech, says that in restrospect he ought to have considered using manual gestures instead, since their chimp, Gua, often attempted to use such gestures, which were of course steadfastly ignored by her speech-oriented trainers. The fact that chimpanzees use brachial and whole-body gestures in the wild, according to Goodall (1971), suggests that this is a much more appropriate vehicle for the study of symbolic capacity in the great apes than speech.

The first attempt to teach a manual language to a chimpanzee was made by the Gardners (1969), with the famous chimpanzee Washoe. It needs to be emphasized that training took place in circumstances very much like that in which a human child learns speech – during everyday activities such as eating, dressing, toilet training, and play. Washoe acquired over 100 signs ultimately, and this record of achievement has been excelled by other chimps subsequently trained by the Gardners and by their collaborator and student, Roger Fouts. Equally important is the fact that the Gardners have established many parallels between the way the chimps learn to sign, including the kinds of mistakes they make, and those of young children (Gardner and Gardner 1971).

Such behaviours, then, constitute a rough parallel to our capacity for language. Consider some of the characteristics: (1) semanticity – there is no doubt that the chimpanzee has the ability to use a sign or symbol to stand for a referent, the simple fact of their signing vocabulary attests to this; (2) arbitrariness – although many signs acquired by the chimpanzee are iconic (e.g., "toothbrush," "drink") many are not (e.g., "dirty," "more"); (3) displacement – several chimps have shown evidence of using a sign in the absence of the referent, and some in the absence of immediate context; (4) productivity – this is more problematic, because productivity would arise frequently only when there is a great deal of spontaneous signing with other conspecifics. It is quite clear that chimps do not spend a lot of time signing to each other. However, a number of instances of the generation of new compositives of signs by the chimp have been reported (Fouts 1973). Washoe generated the amusing expression "dirty Roger" when she was annoyed with her trainer, as well as the composite "water bird" to refer to a swan, where previously the words "water" and "bird" had been used only separately. (5) Transmission to other conspecifics – there is some evidence that a chimpanzee mother has attempted to teach manual signs to her infant, but so far apparently without success.

The conclusion that the chimps use sign language in the way that we do, or rather, as native signers do, has been much criticized. The criticisms focus on the lack of preciseness of the gestures, the fact that they are often employed in a non-communicative situation, and the relatively sparse use of them spontaneously for communication, particularly with other chimps. These extreme positions ignore the fact that the chimpanzee musculature is not precisely like our own, and that chimps apparently do not *need* a manual system of communication to interact with each other, since they have a good facial and whole-body system, and therefore one would not expect them to employ what is essentially a foreign language except under special circumstances. If I suddenly began manually signing when we were communicating quite well in English, you would certainly be puzzled. Finally, many of the criticisms about the non-language use of the signs could be levelled against children as well, who may "babble" in sign before they have become expert in it, much as they babble in speech.

The question as I see it is not whether chimps so far have acquired human levels of expertise, subtlety, and productivity in signing. Of course they have not, and they almost certainly will not. No more will humans acquire the ability to hoot and stamp in an alarming situation exactly as a chimp does. This difference is not, in my opinion, the crux of the significance of work like that of the Gardners and Fouts. The point is that the chimpanzee can employ a rudimentary manual language when constrained to do so. The capacity is there to be selected, as it seems to have been in humans.

It appears, in fact, that the symbolic or semantic capabilities of chimpanzees may be well above anything they normally convey in either man-trained or natural communication. Their symbolic capacity has been studied by Premack (1971) in situations in which nonsense objects, say a blue plastic squiggle, may represent a thing such as an apple, another nonsense object represent the verb "give," and so on. A large array of object-labels is readily learned by the chimpanzee, and in this situation the chimp exhibits a high level of semanticity. Not only can she acquire an object "name" but also concepts like "name of," "negatives," among others. One should say, rather, that this particular training procedure allows us to *see* the level of semanticity; it was there all along but there was no ready way to test it.

Communication in Hominids

If it is painfully difficult to try to establish the intellectual characteristics of animals that continue to exist, such as the great apes, it is obviously even more difficult to do this for organisms that are now extinct. All that remains are structural and physical clues – the kind of skeleton early hominids had, and the kinds of artifacts they left. These are important, but often they only tantalize us with the possible, rather than signify what actually happened. Clues from artifacts are more directly informative about a hominid's abilities, but of course these, too, are incomplete. Stone objects, for example, would more readily survive, bone artifacts would be less likely to survive in a recognizable form, and wood rarely, if at all. Thus the probability that very early man used clubs made of tree limbs is a popular and probably correct inference, but there is no physical evidence for it, because wood seldom leaves a fossil record.

If we consider first the vocal system of communication, what can we deduce about the complexity of speech in early hominids? You will recall that it is now generally agreed that the chimpanzee does not have the requisite vocal tract for generating human speech sounds. Studying the reconstructions of supralaryngeal vocal tracts from early hominids, Philip Lieberman (1975) has concluded that the modern

THE SUPRALARYNGEAL VOCAL TRACT

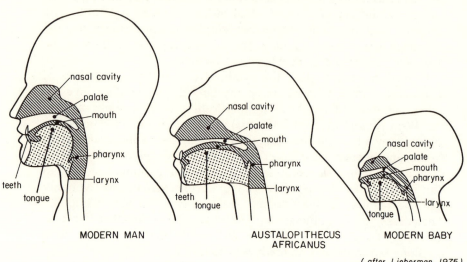

MODERN MAN   AUSTALOPITHECUS   MODERN BABY
                        AFRICANUS

(after Lieberman , 1975)

Figure 1 Schematic representation of variations in the supralaryngeal vocal tract

vocal tract appeared less than half a million years ago. It was not present in Australopithecus, in homo habilis, or even in Neanderthal man. Figure 1 illustrates a simple version of the supralaryngeal vocal tract in an adult human, in Australopithecus, and in an infant human. Only the adult human is capable of making the complex sounds we employ in speech, and note that only the adult human has the right-angled shape to the supralaryngeal vocal tract that appears to be associated with the production of three critical vowels (as in "hod," "heed," and "who'd") in our modern human speech.

While vocal sounds could be emitted, the complexity of speech would be very much less than it is today. The stop consonants (d, b, g, and their unvoiced equivalents) would be especially affected. If correct, these findings suggest that the evolution of complex speech was probably a rather late development, appearing long after the advent of upright walking and of quite complex tool use. Does this mean that only a very crude language existed until 200,000 years ago? I think not.

Table 2 A Summary of Hominid Ancestors

Years before present	Species	Characteristics
8 million	Ramapithecus	Teeth have some similarity with hominids
4 million	Australopithecus afarensis	Completely upright walker; relatively small brain; teeth have some similarity to homo and to africanus
2.5 million	Australopithecus africanus	Upright walker; use of bones as clubs? may not be ancestral to homo
2 million	Homo habilis	Upright walker; tool user
1 million	Homo erectus	Abundant tool use; relatively large brain
1/2 million	Homo sapiens	Modern man

(after Johanson and Edey 1981)

Homo habilis, who first appeared over a million years ago, and who could not only use objects as tools but shape them as well, clearly possessed highly developed manual skill. Indeed, the neural control of the manual and brachial musculature must have been extensive for the tool formation and tool use that are manifest in fossil remains. It is quite possible, therefore, although it cannot be proved, that a manual system of communication was also employed by hominids (Hewes 1973); it may well have been more complex than the vocal system available at the same time. The fact that all modern humans can readily learn a manual language, and that chimpanzees and gorillas can learn at least the rudiments of a manual system of communication, makes it very probable that complex manual systems developed before complex vocal systems in man.

Is there any other evidence that the manual skill known to be present in earlier hominids could have actually mediated a language system? My answer is somewhat indirect, and depends on the fact that the control of speech by the central nervous system, and of manual sign languages in particular, is very closely linked with the system for the control of skilled hand movements. Some of the evidence for this comes from my own research, which I would like to describe briefly (Kimura 1979; 1982).

Central Nervous System Control in Communication

As almost everyone now knows, the two cerebral hemispheres in man are specialized for rather different functions. Injury to or pathology of the left cerebral hemisphere results in a speech disorder, or *aphasia*. This happens only rarely after damage to the right hemisphere. Instead, the right hemisphere appears to be very important for precise perceptual processing and for certain kinds of spatial function. Not surprisingly, the inference from such findings originally was that the left hemisphere contained a system specialized for linguistic function, and this was certainly the assumption made for many years after the early studies of Broca and Wernicke on the effects of left-hemisphere damage (see Geschwind 1972).

However, Broca noted very early on that the left hemisphere was also the hemisphere that controlled the right hand, the hand most of us prefer when asked to do a wide variety of everyday actions. This fact suggests that the left hemisphere may not only govern speech, but that it may have more general motor control functions. Further support for this idea comes from the finding that people who are aphasic from damage to the left hemisphere have difficulty not only in making the movements of speech, but also in making oral movements unconnected with speech; if they have trouble in producing even a single speech sound, they will have trouble performing a non-speech movement on demand, such as moving the tongue to one side. This kind of difficulty is called oral *apraxia*. If their difficulty with speech occurs not with the single syllable, but when producing several syllables, then they will have difficulty carrying out a series of non-verbal oral movements. Their speech problems may then be related to motor control difficulty.

There is even more convincing evidence for the special motor programming function of the left hemisphere: speech disorders are often accompanied not only by oral disorders, but also by manual apraxias. Aphasic persons have great difficulty in copying a series of hand movements that are clearly meaningless or non-representational. We can test such praxic function another way by having the patient learn a task like the one shown in Figure 2. Patients with injury to the left hemisphere have difficulty learning this very simple task, but those with right-hemisphere damage learn it with no problem. I want to emphasize that this is not a difficulty moving the limbs; such patients may have excellent motility, but they cannot produce the correct movements.

Figure 2 Manual sequence box. The task is to learn the three hand postures in the sequence shown in five successive correct trials.

An apraxic person may sometimes be able to perform relatively easy familiar actions, such as brushing his or her teeth, when a particular object is present, but have considerable difficulty when asked to perform such actions in the absence of that object, when required to learn a new motor skill, or when trying to imitate immediately some unfamiliar movements. This suggests that the left hemisphere of the human brain, and especially the parietal and frontal lobes, has some critically important motor programming functions. Indeed, this programming system actually controls both sides of the body, that is, both hands. It must therefore be separate from, and yet operate on, the more direct corticospinal systems, which we know are crossed systems. There is some favouring of the opposite right hand, but the programming system clearly affects both sides. This is a system that would be tremendously important for many everyday non-linguistic manual activities.

The question one might ask is, does this system change in character if the person has learned a manual sign language? That is, are *linguistic* manual movements performed by the same or a different system? The answer seems to be that they are performed by the same system. The example of those who communicated by manual signs rather than speech, and who subsequently became aphasic in their sign language, clearly shows that it is the left cerebral hemisphere that was affected in nearly all reported cases (Kimura 1981). And, in one patient whom I studied personally with the help of signing colleagues (Kimura, Battison, and Lubert 1976), and in at least one other (Chiarello, Knight, and Mandell 1982), the difficulty in producing meaningless movements was just as great as in hearing apraxics who had never signed in their lives. It does not appear, therefore, that the brain organizes linguistic-semantic movements very differently from non-linguistic movements. Linguistic movements simply form a large class of arbitrary learned movements. The implication is that the linguistic component of the movements we make while speaking or signing was probably "added" later to movements already present for some other function.

What kinds of movements could have been present several million years ago that could readily have been adapted for manual communication? The most obvious are the manual abilities involved in weapon and tool use. We know that our hominid forebears have been upright walkers for at least four million years. This means that the arms were free of locomotion and could have been employed in carrying (Lovejoy 1981), throwing (Calvin 1983), chopping, and cutting. Indeed, the variety of stone implements for the latter purpose is impressive as long as two million years ago (Leakey and Lewin 1977), and one assumes that non-stone tools existed before that. Manual signing capacity would have been enhanced if the two hands could have been used relatively independently, and it is clear from the evidence of artifacts that this was in fact the case for a very long time. So we have strong reason to believe that the capacity to make gestural movements with arms and hands has been present for a very long time, perhaps in contrast to the capacity to produce complex speech, which may be relatively recent in origin. That the capacity to attach symbolic function to such manual movements was also available is suggested by its presence in modern chimpanzees and gorillas.

If complex vocal language was indeed a later development, there would have been a motor programming system in the left hemisphere already available to it. The question how and why the vocal system became predominant is interesting but does not yield a definitive answer. There are obvious advantages in a vocal over a manual communication system: discourse could be carried on in the dark and at a distance, and hands would be free to be employed in other tasks while talking.

One might expect, if modern human abilities were very special, that some features of brain organization would be associated with these unique functions. Studying the brain directly through fossil remains is not possible, but significant external features can be studied indirectly by looking an endocasts of the interior of the skull. Holloway (1974) has made a strong case for the position that there are some distinctive characteristics of the human brain that were already present at least two million years ago. These relate to the relative sizes of different functional areas, the occipital (related to vision) being relatively larger in chimpanzees, and the parietal (subserving praxic and spatial function) relatively larger in humans. The latter type of organization according to Holloway is present in Australopithecus as well, suggesting that a human-like brain may have been present before the evolution of complex speech.

Another feature of brain organization once thought to be uniquely human is no longer so considered. This is the presence of functional and anatomical asymmetry. The division of labour between left and right cerebral hemispheres in homo sapiens, outlined earlier, is associated with some known, albeit subtle, anatomical asymmetries. For example, that the superior temporal plane on the left is larger than the corresponding region on the right is reflected in a longer Sylvian fissure on the left than the right (Rubens, Mahowald, and Hutton, 1976). Chimpanzees and orangutans show a similar anatomical asymmetry, though monkeys do not (LeMay and Geschwind 1975; Yeni-Komshian and Benson 1976).

Demonstrating a *functional* asymmetry has so far eluded investigators working with monkeys, probably because the hemisphere has not been adequately defined. Non-human primates apparently do not have a consistent hand preference across individuals, although there is a preference within individuals. If the hemisphere is defined vis-à-vis the preferred paw or hand, rather than as left or right, then functional asymmetries may be evident even in the cat (Webster 1972), and perhaps also in the monkey (Gazzaniga 1963). Similar studies have not been done on the great apes, since this would require doing neurosurgery on expensive and humanly valued animals. It is safe to infer, however, that brain asymmetry does not reflect a specialization of function related specifically to speech or language. That various types of brain asymmetry are present not only among many mammals but also in birds (Nottebohm 1982) is a general phenomenon in the animal world that has yet to be adequately explained.

Conclusions

What I hope I have conveyed is that there is no strong reason to believe that our present abilities are totally different in kind, either from our hominid ancestors, or our great-ape cousins. The continuities in brain structure and in psychological function, and the fact that language does not appear to occupy completely distinct systems in the brain, all suggest that our present intellectual makeup is the product of a very long and slow process of evolution. There may have been intensive periods of change, but abrupt breaks in the history seem unlikely.

If this view is correct, we have to look at some very basic functions to account for our present complex abilities. Such functions are likely to emphasize spatial and computational ability, manual skill, fundamental communicative functions, and the generating of explanations. Wonderful as such abilities are, they present no insuperable difficulties for a non-design explanation.

Research for this paper was supported by grants from the Medical Research Council and the Natural Sciences and Engineering Research Council of Canada.

WORKS CITED

Bennett, J. 1976. *Linguistic Behaviour*. Cambridge: Cambridge University Press.

Calvin, W.H. 1983. "A Stone's Throw and Its Launch Window: Timing Precision and Its Implications for Language and Hominid Brains." *J. Theor. Biol.* 104:121–35.

Chiarello, C., R. Knight, and M. Mandell. 1982. "Aphasia in a Prelingually Deaf Woman." *Brain* 105:29–51.

Chomsky, N. 1967. "The General Properties of Language." In *Brain Mechanisms Underlying Speech and Language*. Ed. T.H. Millikan and F.L. Darley. New York: Grune and Stratton.

Fant, L. 1977. *Sign Language*. Northridge. California: Joyce Media.

Fouts, R.S. 1973. "Talking with Chimpanzees." *Science Year*. 34–49.

Gallup, G.G. 1977. "Self-Recognition in Primates: A Comparataive Approach to the Bidirectional Properties of Consciousness." *Amer. Psychologist* 32:329–38.

Gardner, B.T. and R.A. Gardner. 1971. "Two-Way Communication with an Infant Chimpanzee." In *Behavior of Nonhuman Primates*. Ed. A.M. Schrier and F. Stollnitz. New York: Academic Press.

Gazzaniga, M.S. 1963. "Effects of Commissurotomy on a Preoperatively Learned Visual Discrimination." *Experimental Neurology* 8:14–19.

Geschwind, N. 1972. "Language and the Brain." *Scientific American* 226:76–83.

Goodall, J. van Lawick. 1971. *In the Shadow of Man*. London: Fontana Books.

Hayes, C. 1951. *The Ape in Our House*. New York: Harper.

Hewes, G.W. 1973. "Primate Communication and the Gestural Origin of Language." *Current Anthropology* 14:5–32.

Hockett, C.D. 1960. "The Origin of Speech." *Scientific American* 203:88–96.

Holloway, R.L. 1974. "The Casts of Fossil Hominid Brains." *Scientific American* 231:106–15.

Humphrey, N. 1984. *Consciousness Regained*. Oxford: Oxford University Press.

Johanson, D., and M. Edey. 1981. *Lucy: The Beginnings of Mankind*. New York: Warner Books.

Jürgens, U. 1979. "Neural Control of Vocalization in Nonhuman Primates." In *Neurobiology of Social Communication in Primates*. Ed. H.D. Steklis and M.J. Raleigh. New York: Academic Press. 11–44.

Kellogg, W.N. 1968. "Communication and Language in the Home-Raised Chimpanzee." *Science* 162:423–27.

Kimura, D. 1979. "Neuromotor Mechanisms in the Evolution of Human Communication." In *Neurobiology of Social Communication in Primates*. Ed. H.D. Steklis and M.J. Raleigh. New York: Academic Press. 197–219.

Kimura, D. 1981. "Neural Mechanisms in Manual Signing." *Sign Language Studies* 33:291–312.

Kimura, D. 1982. Left-Hemisphere Control of Oral and Brachial Movements and Their Relation to Communication." *Phil. Trans. Roy. Soc. London* B298:135–49.

Kimura, D., R. Battison, and B. Lubert. 1976. "Impairment of Nonlinguistic Hand Movements in a Deaf Aphasic." *Brain and Language* 3:566–71.

Klima, E., and U. Bellugi. 1979. *The Signs of Language*. Cambridge: Harvard University Press.

Köhler, W. 1959. *The Mentality of Apes*. New York: Vintage Books.

Leakey, R.E., and R. Lewin. 1977. *Origins*. New York: E.P. Dutton.

Lemay, M., and N. Geschwind. 1975. "Hemispheric Differences in the Brains of Great Apes." *Brain, Behavior and Evolution* 11:48–52.

Lieberman, P. 1975. *On the Origins of Language: An Introduction to the Evolution of Human Speech*. New York: Macmillan.

Lovejoy, C.O. 1981. "The Origin of Man." *Science* 211:341–50.

deLuce, J., and H.T. Wilder. 1983. *Language in Primates: Perspectives and Implications*. New York: Springer-Verlag.

Nottebohm, F. 1981. Laterality, seasons and space govern the learning of a motor skill. *Trends in NeuroSciences*, 104–6.

Premack, D. 1971. "Some General Characteristics of a Method for Teaching Language to Organisms That do not Ordinarily Acquire it." In *Cognitive Processes of Nonhuman Primates*. Ed. L.E. Jarrard. New York: Academic Press. 47–82.

Premack, D., and G. Woodruff. 1978. "Does the Chimpanzee Have a Theory of Mind?" *The Behavioral and Brain Sciences* 4:515–26.

Rubens, A.B., M.W. Mahowald, and J.T. Hutton. 1976. "Asymmetry of the Lateral (Sylvian) Fissures in Man." *Neurology* 26:620–24.

Savage-Rumbaugh, E.S., D.M. Rumbaugh, and S. Boysen. 1978. Linguistically Mediated Tool Use and Exchange by Chimpanzees (Pan Troglodytes)." *The Behavioral and Brain Sciences* 4:539–54.

Skinner, B.F. 1976. *About Behaviorism*. New York: Vintage Books.

Stokoe, W.C., D.C. Casterline, and C.G. Croneberg. 1965. *A Dictionary of American Sign Language*. Washington, D.C.: Gallaudet College Press.

Webster, W.G. 1972. "Functional Asymmetry between the Cerebral Hemispheres of the Cat." *Neuropsychologia* 10:75–87.

Yeni-Komshian, G.H., and D.A. Benson. 1976. "Anatomical Study of Cerebral Asymmetry in the Temporal Lobe of Humans, Chimpanzees, and Rhesus Monkeys." *Science* 192:387–9.

More Gaps for God?

It is often argued that the basic laws of the universe seem very finely tuned for the production of life and even of human life. My brief is to comment on these arguments from the point of view of Christian theology.

The subject divides into five topics: (1) the need for arguments supporting Christian belief; (2) the nature of these supporting arguments; (3) the theologically equivocal nature of arguments from "gaps" in the intelligibility of the universe; (4) the possible force of such arguments notwithstanding; and (5) the application of these considerations to the place of miracles within Christianity.

The intellectual world seems to be divided between those atheists who think there is no good reason for believing in God that does not covertly assume what it has to prove, and therefore do not believe in God; and those theists who agree with the atheists that there is no good reason for believing in God, but believe in God just the same – a fashion which may be said to have been set by Sören Kierkegaard and Karl Barth. I do not support either of these views. I agree with the atheists that theism and Christianity *need* supporting arguments to be worthy of belief; but I maintain that such arguments can be provided. I believe that Anthony Flew is right in saying that, short of sound arguments that God exists, there is a presumption that God does not exist. The spectacle of religious persons who admit that they have no good reason for what they claim to believe is almost bound to impel honestly thoughtful people to religious scepticism. Others are apt to resort to the kind of Christianity, or perhaps successor to Christianity, recently publicized by Don Cupitt – where one uses much of the traditional Christian machinery for the kind of spiritual expansion commended by many Eastern religious traditions, without committing oneself to belief in God. It is natural and proper in this situation that people should be excited by new *prima facie* evidence for design in the universe. But, as I shall try to bring out in what follows, I find it very difficult to make up my mind about the theological bearing of this evidence.

The fact is that apologetics is a necessary exercise for intellectually active Christians. They ought, as St Peter says, to be able to give good reason for the faith that is in them.[1] I shall sketch the overall shape that I believe a Christian apologetics ought to take, and then look briefly at these cosmological claims and the arguments based on them. Very briefly, I think that arguments for the existence of God should be primarily epistemological, those for the special claims of Christianity psychological and historical. The argument for God is roughly this. Science both presupposes and confirms that we live in an intelligible world, one that yields to our attempts to investigate it. Any particular hypothesis that is advanced by the scientist is liable to falsification by the evidence; what cannot conceivably be thus falsified, however, is the presumption that one tends to get at the truth about things by propounding and testing hypotheses. A coherent hypothesis is intelligible; so it seems that the world, if it is ultimately found or if it could be found to conform to a set of hypotheses, is an intelligible world. Given that we live in an intelligible world, and that we are not obscurantists, it seems proper to ask what is the best explanation for its being intelligible. It does not appear plausible, or in the long run even coherent, to claim, in the manner of the empiricism descended from Hume, that all that we can know really to exist is the stream of sense-perceptions; or, with Kant, that our minds impose intelligibility on a flux of sensation somehow derived from real things, the actual knowledge of the nature of which remains ineluctably debarred from us. To cut a very long story very short, there is a "fit" between the overall nature and structure of the world and our mental processes; and the best explanation for this "fit" is that the world comes about through something analogous to our minds, which wills the world-order as a whole rather in the same way as we will our actions and products. That the world is intelligible at all is due to the Creator's intellect; that it has the particular sort of intelligibility that it has (that it is to be understood in terms of oxygen rather than phlogiston, of relativity rather than a luminiferous ether) to the Creator's will. Matter in the remotest galaxies, as C.S. Lewis expressed it, must obey the thought-laws of the scientist here in a laboratory, or the whole edifice of science lies in ruins.[2]

My own feeling about this "epistemological argument"[3] is that it puts theism on a much sounder basis than intellectuals have been inclined to attribute to it over the last two centuries. Every educated person "knows" that Hume and Kant between them have put paid to "natural theology," to efforts to argue positively from the existence and overall nature of the world to the existence and nature of a supposed Creator. But at what price! Hume's principles, when consistently carried through, yield a total scepticism about any world supposed to exist prior to and independently of our experience; while Kant's imply that we must despair forever of knowledge of things as they are in themselves. A cardinal principle of both philosophers, that nothing can properly be an object of knowledge that cannot be an object of experience, when thoroughly applied, is just as destructive of physics, history, and knowledge of other minds, as it is of natural theology.

Given that it is so widely assumed that the arguments for natural theology have been destroyed by Hume and Kant, a few more comments on this matter seem in order. Hume's theory of knowledge renders impossible the kinds of knowledge that science and common sense both take for granted; that is, those based upon our experience but supposed to be of a world that exists and is largely as it is prior to and independently of our experience. It is a very odd sort of scientist, and an even odder sort of person of commonsense, who will deny that, while we get to know about trees, dinosaurs, and stars through inquiry based on experience, there were trees, stars, and dinosaurs, with the qualities we find them to have, prior to and independently of this inquiry. But it is just the nature and ground of this knowledge, and the overall nature and structure of the world it progressively reveals, from which the epistemological argument takes its rise. In his attack on cosmological arguments for the existence of God – one of which, significantly, he had defended at an earlier stage in his career[4] – Kant inveighs against the combination of *a priori* and *a posteriori* elements in these arguments. And yet it is just here, on the view I am putting forward, that lies their strength. Kant's great epistemological discovery can be summarized as follows: *a priori*, the world is intelligible; *a posteriori*, we have to determine progressively, by appeal to sense-experience, what sort of intelligibility it has. It is his next move that is the trouble; he claims that since intelligiblity is imposed on things by our minds *a priori*, it cannot belong to things as they are in themselves, which must consequently remain for ever unknown to us. But if the intelligibility does not belong to the real world – in Kant's phrase, to things as they are in themselves – then science is not in the business of finding out what is really the case about real things, and indeed any kind of objective knowledge is impossible. If it does, then the question of God can be clearly asked and affirmatively answered; the cosmos is permeable to our minds because something in its constitution is analogous to our minds. The pioneer scientists of the seventeenth century, as Ilya Prigogine has recently emphasized, took for granted that, in Kepler's words, they were thinking Gods's thoughts after God; and if the way of thinking I am recommending is correct, they were justified in doing so.

It may well be asked whether appeal in this context to something like what many have called God is not a matter of foreclosing the search for scientific explanation, whether as a result of obscurantism or of sheer exhaustion. This view seems to me to overlook what one might call the second-order nature of the question asked and the answer given to it. The question, "of what overall nature and structure must the universe be to be susceptible to scientific investigation and explanation at all?" can hardly be itself a scientific question in quite the usual sense. To insist that answers to such questions be "naturalistic" rather than "supernaturalistic" may amount to little more than to insist that one keeps God out of it at all costs. To say that the question is an "unreal" one is to run into the objection that so many thinkers have thought it worth asking and answering in some fashion – as when one claims that the implications of science are mechanistic or involve dialectical materialism or phenomenalism. The fashion now seems to be to follow Fritjof Capra in saying that the metaphysical implications of physics are a sort of oriental mysticism; such representatives of nineteenth-century scientific materialism as Büchner and Haeckel must be turning in their graves. Not to ask the qustion at all might be held to be no less obscurantist than to put up too readily with invocations of the deity at the first-order level, where, to be sure, one ought always to look for and expect natural explanations – a fact whose implications we shall have to consider later on. As Daniel Goldstick recently remarked in debate, to say that such questions are "fruitless" might really amount to little more than a protest that one did not like the fruit they yielded.

Are the asking and answering of this question in any way scientific? As I have just argued, questions of this sort are of too great a generality to be scientific in a strict sense. But one might urge their legitimacy in view of the fact that they seem quite closely analogous to scientific questions and answers in the usual sense. One asks, first, of what overall nature and structure the universe must be in order to be susceptible of scientific explanation; and, second, what is the best way of accounting for its having this nature and structure. In science, at least as often described, one thinks up a range of explanations for a phenomenon and retains provisionally the one that is most satisfactory. Rather similarly, in the case of the phenomenon of the intelligibility of the world, a number of more-or-less plausible explanations (that we impose the intelligible structure on the given flux of phenomena, that things in themselves are unknowable, that the question ought not to be asked) may be proposed and rejected as inadequate, and the preferred one – that its intelligibility is due to creative intelligence – accepted as the most satisfactory available.

I suppose that a typical exponent of a non-theistic evolutionary world-view might object to the train of thought I have outlined somewhat as follows: "Thought is not to be regarded as self-sufficient and self-justifying, but rather in the context of biological evolution. Over billions of years, by a process of mutation and natural selection, life, animal life, and intelligent animal life, have successively evolved. Rationality is of survival value; given enough time it is likely to occur and to prosper in a universe such as ours. In the struggle of humanity for survival, in relation to the non-human environment and to other human groups, the capacities to observe, hypothesize, infer, and test, which in the long run issue in the imposing edifice of the sciences, are at a premium. On this account of things, no *a priori* fit is required between the things and processes that constitute the material world on the one hand, and the operations of the human mind on the other."

It seems to me that this kind of explanation for the intelligibility of the world places the cart firmly before the horse. It is to be noted in the first place that what is at issue is not merely successful interaction with an environment, but knowledge of a world; while the latter evidently presupposes the former, it is by no means to be identified with it. Let us concede for the purpose of argument – what Professor Swinburne provides us with impressive reasons for *not* conceding – that evolutionary theory in its usual form is adequate to account for the ultimate emergence of consciousness from matter. The question still remains, of how there came to be an intelligible world in the first place, before minds came into being within it such as were capable of actually understanding it. It is all very well saying, in the manner of Hume's Epicurean,[5] that matter in its vicissitudes over virtually infinite time was bound or at least likely to get into that enormously complicated and relatively stable state that supports or is identical with the life of the mind. But what *is* such "matter"? Either it is the realization of some intelligible possibility, or it is not. If it is, it is itself an aspect of that general state of affairs that is only ultimately to be accounted for, if I am right, by the postulation of something like an intelligent will at the basis of things – the intelligence ensuring their intelligibility, the will just what kind of intelligibility they have. If it is not, it is a mere metaphysical excrescence, the ghost of a naive realism surviving incongruously within the scientific world-view – what Eddington called "matter (Loud Cheers!)" as opposed to matter in any sense essential to science. I take it that mechanistic materialism has imposed itself on the minds of many scientists as a mistaken apprehension of the truth that the universe must be intelligible. I would argue that it is this mechanistic materialism, and not the success of the scientific enterprise as such, that has led to that disenchantment of the world deplored by so many thinkers from Max Weber to Ilya Prigogine. Impressive in many ways as they are, revivals of mechanistic materialism by such analytical philosophers as J.J.C. Smart and D.M. Armstrong seem to me like the spasms of a dying animal.

When it comes to the special claims of Christianity, given the acceptance of theism, another kind of argument is in order. The work of Jesus Christ is supposed to constitute a remedy for the disorder in individuals and society that Christians call sin. Two obvious questions then arise: does the alleged remedy seem to meet the disorder; and is there evidence that it has actually been provided? For better or for worse, Christianity has a great deal in common with other religions in which human beings participate, by ritual and recitation, in the fate of a cult-hero. Jesus, in common with other cult-heroes, has a mysterious birth, is threatened soon afterwards with deadly dangers, is rescued in the nick of time, has a precocious development, performs miracles, and has a tragic and early death along with a symbolic manner of dying followed by remarkable post-mortem effects.[6] Jesus's career as reported to us in the Gospels, in other words, is a very complete version of the archetypal myth found in more or less fragmentary form all over the world. By identification with Jesus, Christians hope to canalize and transform their affectivity and aggressivity in the service of the good, and so to overcome sin and its results in themselves and in the world. Konrad Lorenz wrote that, if human aggression is to be controlled, four things above all are necessary: a cause, a community, an enemy, and a leader.[7] As Christians see it, there has been providentially given the one leader able and worthy to carry the enormous ideological weight that human beings place on their leaders, the man who is also God; to lead the community, the Church, which cuts national and other group loyalties down to size; for the cause, which is goodness and truth as such; against the enemy, which is evil as such, and not the human beings who are its instruments, victims, or dupes. How far Christianity may be effective in this respect is a matter of psychological argument. But even granted that it is so effective, there remains the question of how far it is literally true, historically true; unless the events described in the Gospels, or at least something very like them, *really happened*, Christianity is just one system of myths among others. To some extent, then, the Gospels, however profound they are just as stories, must turn out to be historically true, when subjected to critical investigation, if Christianity is to be true. To parody Kant's famous remark about concepts and intuitions, the myth without the history would be empty, whereas the historical content without the mythical form would be pointless. The essence of Christianity, to quote C.S. Lewis again, is that here and here alone (at least so radically) "myth became fact":[8] there was a real God-man who really lived the archetypal myth, doing for humankind what humankind has tried to do for itself in the profound but literally false "histories" and cosmologies that are "myth" in the usual sense.

If these epistemological, psychological, and historical arguments are valid – and of course they are each enormously contentious – there is a good case for Christianity; and one can agree with the following comment of A.R. Peacocke: "The collapse of the medieval theological citadel during the last three hundred years and its recent undermining from within has tended to hide a soundness in its foundations which is greater than is usually conceded by those whose education has lacked any disciplined inquiry into the intellectual development of Christianity."[9]

I have said that Christianity needs an apologetic (what is affirmed gratuitously, as the scholastic tag has it, is denied gratuitously), and have sketched the general form I believe such an apologetic should take. (I should add in passing that the soundness of the arguments used, since on any showing they fall short of proof, by no means removes the need for faith, except for that immoral and irrational kind of "faith" that stifles or contradicts reason rather than promoting and supplementing it.) But what I want to emphasize is this. *Where theism at least is concerned, the stress in apologetics as I have sketched it is on explanation for the intelligibility of the world, rather than on accounting for the gaps in that intelligibility.* From this perspective, one may be inclined to wonder whether additional arguments for the existence of God based on *gaps* in the intelligible fabric of the world may not rather be in the nature of an embarrassment than a help. Now it is sometimes disputed whether God should be treated as an hypothesis. But there are at least two senses in which the claim that God is (or is not) a hypothesis may be understood. That God is not a hypothesis may imply that God is not required for the explanation of anything whatever. But if this is so, it seems to me, as it seems to the atheists whom I mentioned at the beginnning of the paper, that one might as well deny that God really exists at all, except in a rather trivial sense as an intentional object of religious discourse and devotion. (In this trivial sense, "God" exists just so long as there are practitioners of theistic religion.)[10] But that God is not a hypothesis may imply only that, while God may be invoked as explanation for those general characteristics of the world that ensue from the fact that it is knowable at all, God is not to be invoked as explaining particular states of affairs within it.

Hume's *Dialogues Concerning Natural Religion* give classical expression to a dilemma that is apt to bother reflective theists, and to be pressed by their atheist and agnostic opponents. If one invokes God as explanation of particular examples of *prima facie* contrivance within the cosmos, one may seem to be dragging God down to the human level, thus falling foul of the heresy of anthropomorphism. But if one is so fastidious in this respect as to leave God no explanatory function whatsoever, one is likely to render the supposition that God exists at all meaningless or otiose. One of the charms of the epistemological argument is that it enables one to provide a clear solution to the problem posed by this dilemma. God is like a human being in being an intelligent will that accounts for the whole universe much as human beings do for the states of affairs for which they are responsible, but not like a human being in having to tinker with the results of plans initially made. Clearly evidence for the fine tuning of the initial conditions and fundamental laws underlying the cosmos will give comfort to those who are more harassed by the fear that theism might be reduced to meaninglessness than by the spectre of anthropomorphism. But if a theist is impressed by the epistemological argument, she will be convinced in any case that God has what one might call a general explanatory function, and may be rather embarrassed than encouraged by strong evidence that God has particular explanatory functions as well. If God is the primary cause, ought God to be too readily invoked as a secondary cause? To quote Peacocke once again, the scientific perspective on the history of life and of the cosmos is "of a beauty and an all-embracing scope that, when fully apprehended, dazzles the mind and heightens the sensibilities ... A notable aspect of this picture is the seamless character of the web which has been spun on the loom of time: the process is continuous from beginning to end and at no point does the modern natural scientist have to invoke any non-natural cases to explain what he observes or infers about the past."[11] The deity who conceived and willed this "seamless web" may well be felt to have earned Anselm's praise of being "that than which a greater cannot be conceived."[12] Would God be so much so in balancing things on the knife-edge of extreme improbabilities? I do not assert outright that God would *not* be so much so; it is a matter on which I find it extremely difficult to make up my mind.

I have aired my theological worries about these claims and the arguments based on them. But to admit such worries is by no means necessarily to deny that as arguments to intelligent design in the universe they may have some force. It is of considerable interest that some cosmologists have been led to postulate the existence of an enormous number of quasi-universes of which this quasi-universe happened by chance to be that one with those extraordinary conditions necessary for life to evolve.[13] The facts in question show that the chances of our own quasi-universe coming into existence are billions to one against; the explanation must be, according to these cosmologists, that there really exist the billions or so of quasi-universes that would make the existence of our own quasi-universe reasonably likely as a result of chance.

Now theists are often teased by their atheist and sceptical opponents for failing to apply the principle called Ockham's razor with sufficient thoroughness. Where natural causes are sufficient in explanation – and who can ever confidently and with good reason declare in any case that they are not? – a supernatural cause is redundant. The usual formulation of Ockham's razor is that entities ought not to be multiplied beyond necessity. This is not, of course, without ambiguity; an explanation that literally postulates the smallest number of entities may not be the most elegant or, in an obvious sense at least, the most plausible, of those available in any particular case. (Suppose a burglary has been committed, which was either the work of one individual of exceptional ingenuity and physical strength, or of two individuals of average endowment; to which explanation does Ockham's razor give preference?) But however one interprets the principle, when it comes to the case before us the theist may well feel that the tables are well and truly turned in his favour. Is not a single intelligent will that conceives and wills these singular features of the world-order a far more economical explanation of them than a vast number of other "worlds" in whose existence we have no other reason to believe?

The multiplicity of worlds or quasi-universes may be envisaged to exist either simultaneously or in succession. Given the truth of the "big bang" cosmology, one might think of expansions followed by contractions, "big bangs" followed by big crunches, going on more or less indefinitely. I understand that whether the universe is oscillating in this fashion is still in dispute among those in a position fruitfully to speculate about such matters. However, if it is oscillating, it appears to follow that the galaxies are receding from one another at less than escape velocity, so that the expansion of the universe may in time be halted and reversed by gravity. But I understand that, given the not unreasonable assumption that the large majority of the matter within the universe is associated with the galaxies, the universe is nowhere near dense enough for this to happen.[14] So there does seem to be positive evidence against that version of the "many worlds" theory that has them juxtaposed in time rather than in space.

I earlier expressed my doubts as to whether evidence for gaps in the intelligible fabric of the cosmos is more a help than an embarrassment to the Christian theologian. If the latter, should the role traditionally accorded to miracles in revelation be rejected? Given the place of accounts of miracles in the gospels as we have them, it is difficult to see how one could reject their occurrence or significance while remaining a Christian in any very useful sense. Whoever it was that described the Gospel of Mark as just a succession of miracle stories with a passion narrative tacked on the end was no doubt exaggerating a little, but she had a point. It seems to me that this problem could be coped with in one of two ways. First, one might claim that infringements of laws of nature were appropriate in the special case of signifying the Creator's entry into his own world. To do so would accord with Leibniz's suggestion that miracles belong in the order of grace rather than of nature. Second, and this is the approach with which I have most sympathy myself, one might deny that miracles did in fact constitute gaps in the order of nature in the relevant sense. I believe that there is a considerable amount of evidence that paranormal events (to use a deliberately neutral term) are commonly associated with exceptional spiritual development, whether for good or for evil. On this assumption paranormal events would rather be indications of laws of nature that came into operation in special circumstances than infringements of laws of nature. Rather as (to use Hume's example employed in his discussion of miracles)[15] the notion of water becoming solid might seem incredible to someone brought up in the tropics, but quite to be expected by one who lives farther north, so events that might seem unbelievable to oilmen, academics, and businessmen might become the most "natural" thing in the world when you keep the company of saints, satanists, shamans, or witch-doctors.

NOTES

1 1 Peter 3:15.
2 C.S. Lewis, *They Asked for a Paper* (London: Bles, 1962), 162.
3 Virginia Tumacz provided this useful expression.
4 In the *Dissertation* of 1770.
5 David Hume, *Dialogues Concerning Natural Religion*, VIII.
6 Cf. C.G. Jung, *Psychology and Religion* (London: Routledge and Kegan Paul, 1958), 154ff.
7 K. Lorenz, *On Aggression* (London: Methuen, 1966), 234ff. Lorenz himself does not make the application to Christianity.
8 Lewis, *Undeceptions*, ed. Walter Hooper (London: Bles, 1971), 39.
9 A.R. Peacocke, *Science and the Christian Experiment* (London: Oxford University Press, 1971), 3–4.
10 As has been argued to follow from the philosophical principles of D.Z. Phillips. Cf. *The Concept of Prayer* (London: Routledge and Kegan Paul, 1965).
11 Peacocke, *Creation and the World of Science* (Oxford: Clarendon Press, 1979), 60.
12 Anselm, *Proslogion*, 2.
13 Of course there is a sense in which, by definition, there can only be one universe; I use the inelegant term "quasi-universe" to forestall the obvious objection arising from this fact. The other quasi-universes would all be part of the one universe, in the sense of "universe" that means "the sum-total of what exists"; but causally they would interact little, if at all, with our own quasi-universe.
14 For this contention, see W.L. Craig, *The Kalam Cosmological Argument* (London: Macmillan, 1979), 122. On the new cosmological speculations and their application to theism, I have learned an enormous amount from the articles and conversation of John Leslie.
15 Hume, *Enquiry Concerning Human Understanding*, x.

RAVI RAVINDRA

In the Beginning Is the Dance of Love

It is a reflection of our collective world-view, perhaps since the publication of Newton's *Principia* in 1687, that we regard questions concerning the origin, development, measure, and meaning of the cosmos as pertaining almost exclusively to the domain of science, and in particular to that of physics. In other words, for us moderns, cosmology is a branch of physics, a subject that since the sixteenth century has concerned itself with understanding the cosmos ultimately in terms of dead matter in motion in reaction to external and purposeless forces.

As is well known, natural theology has a long history. At the beginning of modern science, Kepler regarded himself as a priest of God in the temple of Nature. And for Newton, all his scientific work was a gloria in praise of God. Since his time scientists have felt increasingly uneasy about mentioning God, at least in their scientific publications. A long and hard struggle was necessary to establish natural science as an independent mode of inquiry, free of the tyranny of theology and the church, which had been coupled with temporal power. Now, especially since the making of the atomic bomb in 1945, it is science that is associated with power; and a similar struggle may be necessary to rescue genuine spiritual inquiry from the tyranny of scientific rationality.

Contemplation of the heavens has always played a significant role in bringing human beings to wonder about the meaning and purpose of the cosmos and of their own existence. The heavens have always seemed to be the abode of the sacred, inspiring reflection and awe.

However, a subtle shift has taken place in our attitudes, owing to the rise and development of modern science. Let us take a familiar example from Psalm 8. The psalmist asks: "When I consider thy heavens, the work of thy fingers, the moon and the stars, which thou hast ordained; What is man, that thou art mindful of him?" We too have contemplated the heavens and other things in the light of the latest scientific knowledge, but our attitudes, and our questions, are different. If I may be permitted a modern rendering of the psalm, the scientist is more likely to ask: "When I consider the heavens, the work of our equations, the blackholes and the white dwarfs, which we have ordained; What is God, that we are mindful of him?"

Ideas and activities flourish in the context of a world-view, although world-views themselves are permeable and elastic, and change. It is science that is the major component and the strongest constituent of the present paradigm, and it is in the assurance of a shared scientific rationality that all our intellectual discussions now take place.

I do not have any new data and I do not really believe that what we need, collectively or individually, is additional data to come to a proper sense of a design or its absence in the cosmos, and of our relationship with it. What I propose to do is to raise some questions and make some comments about and around the theme of this volume, organizing my discussion under the subheadings conveniently provided by the key words and associated ideas of the title: origin, evolution, universe, evidence, and design. These terms conform to a particular kind of rationality and circumscribe our deliberations here.

Origin

The question of the origin of the universe is intimately connected with the undertanding of time. It is practically impossible for the Western mind, particularly since Augustine in the fourth-fifth century, not to think of time linearly. The notion of linear time has entered deeply into the structure of scientific thinking. Even when we think of non-linear time, as we sometimes do in contemporary physics, we look at the non-classical properties of time: what its conjugate variables are, how it works in other dimensions and spaces, etc. But what is important to emphasize is that in physics we are always dealing with some dimension of time, and never the sort of situation when "time shall be no longer" as is said in Revelation 10:6. Of course, when we extrapolate along the dimension of time, we might run into a singularity, as we do for example in equations dealing with gravitational collapse or the cosmological solutions leading to the "big bang" theory of the origin of the universe. There our notions of time go awry, and we need some very ingenious methods to get around these difficulties. The thing to note, however, is that from the point of view of the physical cosmologists, the questions concerning the beginning of the universe have entirely to do with smaller and smaller amounts of time from the initial event when all this began. However many theoretical or practical difficulties we might encounter, what in fact we are trying to do is to follow the time co-ordinate back to zero. We have theories now dealing with the state of the universe at time spans of the order of 10^{-23} seconds after the absolute zero of time. There are theoretical reasons for believing that this may be the closest we can get to the absolute beginning along the time co-ordinate. What that in fact means is that according to our present notions of time it makes no sense to talk about time any closer to that beginning and certainly not prior to it.

What I should like to suggest, however, is that the sort of beginning that the physical cosmologists search for is not the beginning spoken of in the mystical or the mythical literature. When it is said, for example, in the opening lines of Genesis, "In the beginning God created the heaven and the earth," we are tempted to think that according to the Bible the heaven and the earth were the first manifestations. To do so, however, is a mistake, as we see from the immediately following verses. The heavens were not created until the second day and the earth not until the third; and the heavens, also called the firmament, were created in order to divide the waters above from the waters below. These waters, one should notice, existed before the existence of the heavens and the earth, which, on the other hand, were said to be created in the beginning. Perhaps we are presented with two different kinds of heavens and two different kinds of earth. I shall not engage in biblical exegesis here; all I wish to suggest is that we have here a notion of a beginning that is different from the scientific notion.

Of course we may here be encountering difficulties with language that are endemic to all religious literature, often also to poetry as well as to almost any situation of intimacy. But there is no reason for us to imagine that the scriptures are meant to be at our service and that they must be clear to us while we remain as we are. I imagine that, at the least, scriptures summon us to realities that we do not ordinarily perceive. There is universal agreement among all spiritual traditions that for us to perceive these hidden realities something in us needs to change. We cannot remain as we are and come to the Mystery. That change is called by many names: a change in the level of being, a change in consciousness, a deepening of faith, a new birth, the opening of the third eye, the eye of true gnosis, and so on. One of the fundamental changes that is said to occur when the doors of perception are cleansed concerns time: not only does one's sense of duration change but, more important, what alters radically is one's relationship with the passage of time.

Statements like "In the beginning was the Word," in spite of their appearances, are not statements concerning ordinary time, the sort of time on a co-ordinate axis whose point of origin is the beginning. These statements carry weight and significance precisely because they were uttered and received in heightened states of awareness. It is true that scriptures can be and have been misused to cover up intellectual laziness, to foster fear, hatred, bigotry, and the like. Such misuse can make a spiritual document or symbol fearful and even hateful to people of goodwill. Nevertheless, whenever these writings and symbols speak to anyone spiritually, it is because they carry a higher level of energy and not primarily, or even at all, because of any logical clarity or agreement with our scientific notions of space and time.

This other kind of time, that of myth and mystical writing, is certainly not contradictory to our ordinary time. Nor is it, however, merely an extension of it – in either direction of the time co-ordinate, to the beginning or to the end. Just as the scriptural beginning is not the zero of the time co-ordinate, mystical eternity is not an infinite extension of time. Thus what is everlasting is not necessarily eternal. It appears that spiritual time is in a way orthogonal to scientific time, in the sense that the mathematicians use the notion of orthogonality, which is to say that it lies in a dimension wholly independent of the domain of time, although it is able to intersect with time at any moment. Thus, even if we were to consider time multidimensionally, or even non-sequentially, or any other way, no manipulations of time or in time could lead to the dimension of eternity that is orthogonal to it, and in which mythic beginnings and endings are spoken.

Evolution

As long as there is time there is change. That is how we understand and measure time; that is how we know that time exists and that it passes. It is only in this minimal physical sense, of state a changing into state b, that we speak about the evolution of the universe in physical cosmology. But there is an ordinary use of the word "evolution" that has a certain emotional connotation of which we need to be careful;

otherwise we only introduce a philosophical problem where we do not intend to. Ordinarily, one thinks of evolution as containing within it an idea of change in a desirable direction, so that the end product is at a level higher than the antecedents. Now, it is very difficult to say in what sense one understands "level." There are, however, also connected notions like development, growth, progress, and the like. Something or someone who is at a higher level may have more being, more consciousness, more wisdom, or the potentiality to perform more complicated tasks than one at a lower level. What is important is that the idea of hierarchy is built right into the notion of levels and of evolution, and that furthermore we specify from our point of view which is higher or lower, or whether a process is degenerative, progressive, or static. What cosmologists really talk about is physical change, without attaching any notion of hierarchy of being. But once the universe unfolds what is seen may be judged from a particular point of view. As in the Genesis account of creation, at the end of each day and each new manifestation, God looked at it and pronounced it to be good. Our cosmologists consider what they think happened and pronounce the change to be evolution. All that our physical laws describe is change in time. There is nowhere any place in them for intention, purpose, or evolution as long as it contains the emotionally laden sense of progress in it.

It is worth paying a little more attention to this point. In the history of natural philosophy, ideas relating to change and the laws governing the dynamics of nature have been intimately connected with the notions of causality, and for obvious reasons. Three distinct notions of causality can be distinguished for our purposes here: metaphysical, physical, and biological.

The metaphysical notion of causality, which prevailed until the sixteenth century, assumes that the cause is greater than the effect. Thus, in theology, the creator is naturally greater than the creation, at a higher level of being, intelligence, and power. This principle was applied also in natural philosophy and was, from the point of view of the subsequent developments in science, a stumbling block to a proper understanding of nature.

During the sixteenth century, a new understanding of physical causality emerged, according to which the cause and the effect were at the same level. This was a time in history when, in my view, there was a general levelling off in every field of human culture and society. Now in natural philosophy one did not speak of a cause being higher than, or in some senses containing, the effect. Instead, one spoke of change,

change of one state of matter into another, without raising or lowering its level of being or intelligence or desirability. It was a subtle shift from the domain of intentions, will, reasons, and purposes, and the forces and laws or, in another language, angels and powers, required to carry out these intentions in nature, to a field of forces and laws operating in nature without any purpose. (The philosophical and theological controversy between Leibniz and Newton was connected with this shift in the understanding of causality and the consequent sundering of the realm of facts from the realm of intentions, or of nature from spirit.)

In the nineteenth century a biological notion of causality emerged, according to which the cause is lower than the effect. That which is inferior, ontologically or in intelligence or in the subtlety of cellular organization, gives rise to what is superior. Thus amoebae would give rise, in time, to Einstein. Since what follows is more desirable than the antecedent, from the human point of view, this notion of causality is rightly called evolution. This principle is the inverse of the metaphysical and the theological notion of causality: rather than proceeding from above, creation, including human beings, now proceeds from below. In its wake this idea naturally brings an immense amount of anxiety and unease, especially to those who are comforted by a belief in some ultimate cause, or God, who is personally concerned about their welfare.

Returning now to scientific cosmology, it was only sixty years ago that the idea of the entire universe itself being dynamic was formulated precisely. One of the solutions to the field equations of General Relativity demanded that the universe as a whole be dynamic; otherwise the solution was unstable. This notion of the dynamism of the cosmos seems to have been such a revolutionary idea in the Judaeo-Christian world that even a radical thinker like Einstein balked at it. He tinkered with his equations and introduced another factor into them called the cosmical constant, which was helpful in obtaining a stationary solution to the field equations. Soon after, it was discovered that even with this new, somewhat arbitrarily introduced, constant, dynamic solutions of the equations still resulted. Also, within a few years, Hubble discovered from observational data that the galaxies were receding from each other at the speed of light and that the universe was therefore expanding. This was the most significant observational confirmation of Einstein's theory of General Relativity, and Einstein himself later remarked that the introduction of the cosmical constant in his field equations was "the greatest blunder of

[his] life." The point of these remarks here is that the fundamental equations on which modern physical cosmology is based have nothing to do with evolution, except in the minimal physical sense of change. Modern cosmology is just like the rest of physics as far as the notion of causality is concerned: it describes the change in matter-energy from one state to another. Naturally, from our point of view, the emergence of the stars, galaxies, the solar system, and ultimately of ourselves is more desirable than their non-emergence, and so we feel justified in describing this change as evolution.

What we need to be aware of is that in this process we are combining two different notions of causality described earlier. One of these we actually need for our knowledge; the other is an emotional overlay for the obvious reason that we humans are at the end of the corresponding change. So we get saddled with a philosophical problem because of our sentimentality about human beings while nevertheless insisting on a limited physico-biological view of man. We do not need so limited a view of cosmology that the deepest, spiritual part of ourselves, cannot be taken into account. In physical cosmology, which is a perfectly legitimate and wonderful study in its domain, it is change in the physical form of matter-energy that is our concern; we do not speak, indeed we cannot speak within the assumptions and pro-cedures that govern the subject, of spiritual evolution. Of course, human beings have always had a need and a sense of the sacred; this alone gives meaning and purpose both to ourselves and to the cosmos. Fundamentally bereft of the sacred we are riddled with personal anxiety and adrift in the meaningless vastness of spacetime. Physical theories concerning the static or the dynamic nature of the universe are not, nor do they pretend to be, about the dimension of significance or purpose.

Universe

What do we mean by universe? Presumably, all there is. Does a cat or a bee have the same universe as a man? Does a tone-deaf or a colour-blind person have the same universe as the one who is musically gifted or is a painter? Does a person who is blind to symbols or to spirit, or who is insensitive to wonder, beauty, or spiritual presence have the same universe as a scientist or a poet or a mystic?

What there is is a function of who sees. This axiom is not meant to support the philosophical position that claims that a thing does not exist unless there is someone to see it. My concern is our knowledge:

what we *know*, actually and potentially, about the universe depends on
the procedures, methods, and interests that we bring to our observation of it. If we do not know how to find angels and we are not
interested in them, we will say that the angels do not exist. And it is
true that they are not a part of our scientific universe. Nor are "the
clouds which brood," which were a part of Wordsworth's universe,
nor are the dancing colours inhabiting Blake's universe, nor are the
cherubim and the seraphim singing "Holy, holy, holy" who were a part
of Bach's universe. The physical cosmologist's universe, vast and
marvellous as it is, is not all there is. As Shakespeare would have put it,
"There are more things in heaven and earth than are dreamt of in your
philosophy."

Even when allowance has been made for error and illusion, which
can, of course, as much blight the cosmologist as the poet, the musician,
or the mystic, it is difficult simply to dismiss these other fields. Hardly
anyone of sound judgment and goodwill dismisses the arts out of
hand. But it is astonishing how so many people find it much easier to
dismiss the mystic and the theologian. There are understandable
historical reasons for this, but what concerns us here is that in
intellectual circles none of these fields is now considered as relevant to
deliberations concerning the cosmos. Witness, for example, the contents of this volume. The universes inhabited by what is regarded as
the most precious by the artist, or the musician, or the mystic are
somehow relegated by us to a murky and imaginary realm, not entirely
real. And certainly not as real as the multiple universes or the shadow
universe or the anti-universe or the other weird universes that make up
the speculations of physical cosmologists.

The important point is that it is our assumption now that whatever
else the musicians, artists, or mystics might be doing, they are certainly
not producing knowledge. Knowledge is produced exclusively by
scientists, we would say, and by nobody else. And contemporary
philosophers, with all their love for wisdom, in general agree. We
might not now say, with the positivists, that "non-science is nonsense," but we would surely say that non-science cannot lead to
knowledge and truth.

In connection with what we include in the universe, mention needs
to be made of a traditional idea of levels of materiality. As is well
known, medieval philosophers in general held that the matter on
different planets was different, as were the laws in operation there. It
was a considerable advance in astronomy to establish that fundamentally the same sort of matter prevailed throughout the universe, subject

to the same laws everywhere. However, when we move from medieval natural philosophy, whether expressed in alchemy, astrology, mathematics, or cosmology, to the modern sciences, our general reaction to the backward-looking nature of the past and our excitement over new discoveries blind most of us to the predominantly symbolic and analogical nature of medieval thought. We would do well to remind ourselves of the ancient analogy between each human being and the universe, between the cosmos and the microcosmos that inwardly mirror one another's essential principles. We might then realize that the various planets, the different materials on them, and the different laws operating there were all symbols of different levels of interiority within a human being, and that the quality of matter-energy at different levels of the mind is different from the matter-energy of the body and subject to different laws. Sometimes this idea was explicitly shown in various diagrams, but the prevalence of symbolic and analogical ways of thinking, meant that it was often just assumed, much as we today assume that everyone in all reasonable gatherings naturally accepts the mode of scientific rationality. It is plain and obvious, as Blake succinctly put it, "Reason and Newton are quite two things." What goes on in our minds and our feelings, and not only what takes place in our bodies, also contributes to all there is.

By bringing in mental and psychic functions, I do not wish to suggest that these are in principle outside the domain of scientific knowledge. I am not proposing anything supernatural, as opposed to natural, that is excluded from the investigations of natural philosophy. There is nothing supernatural about most of what gets labelled extrasensory perception, or miraculous. These, to be sure, are at present extra-science perceptions, but there is nothing inherently beyond nature or beyond science in them. It may well be that a radically altered science will be required to understand what is now extrasensory perception, just as a radically altered science was required to understand lightning in the sky or the light of the sun, which might have seemed quite supernatural from the perspective of fourteenth-century scientists. It is important to distinguish, as St Augustine did, between what we claim really is nature and what we know of nature. The limits of our knowledge are not necessarily the limits of nature.

But it is still more important to realize that even with a radically altered science that could take account of extrasensory perceptions and other miraculous happenings, we cannot come to the end of all there is. All there is far exceeds the realm of nature, the domain of causality and materiality, however subtle our descriptions. To say that we do not yet know certain levels of nature is not to say that nature is all that there is to know or that can be. In fact, practically without exception, all great spiritual teachers, such as the Buddha, the Christ, Patanjali, Krishna, and Moses, have warned against an excessive fascination with miraculous phenomena and occult powers, which are said to be diversions from the true spiritual paths.

Two related, although somewhat parenthetical, remarks may be made here. One of them concerns an important distinction, made in the scientific revolution starting in the sixteenth century, between the primary and the secondary qualities of matter. This distinction played a crucial role in the development of the physical sciences, and also in the subsequent impoverishment of nature. The primary qualities were extension, mass, and velocity; to this list was added charge in the nineteenth century and spin, strangeness, charm, and others in the twentieth. The secondary qualities consisted of taste, colour, smell, and the like; they were not considered objectively to be a part of nature, but were subjective and rather unreliable. Considered even more subjective and unreliable were tertiary qualities, feelings of beauty, purpose, or significance. The secondary and tertiary qualities were gradually eliminated not only as instruments of inquiry into nature but also as fundamental constituents of nature. They could not, properly speaking, be studied as themselves constituting reality, but as something which needed to be explained and understood in terms of the primary qualities. Thus a deep-seated reductionism is built into the fundamental presuppositions of scientific inquiry. A division into *res extensa* and *res cogitans* carried within it a certain instability attached to the realm of the mind. From a scientific point of view, as we see clearly in behavioural psychology, all psychic functions must be reducible to external motions. On the other hand, we have the philosophical problem of mind-body dualism. In some theological circles it is really understood as soul-body dualism, in which the soul is supernatural, removed from the realm of nature and scientific investigation altogether, and placed in the realm of faith away from knowledge. Any real knowledge of the psyche or the soul thus gets rather short shrift: the scientists deny the existence of anything in it which they cannot study by physical means, and the theologians deny the possibility of any knowledge of it. But in neither case can spiritual qualities have any independent existence in the cosmos that we can study.

The other related remark derives fom a comparative study of the history of ideas in the Western world and in India. In Greek philosophy, and in the early Christian writers as well as in the Indian tradition, there was a tripartite division of a human being into spirit, soul, and body, or, to use the terminology of St Paul, *pneuma*, *psyche*, and *soma*.[1] Gradually this three-fold division shrank into a two-fold division: spirit and nature, or mind and matter, or soul and body. The *coup de grâce* was dealt by Descartes, who explicitly identifies spirit with soul and both with the mind.[2] In the Western world, since the time of Descartes, soul is in general regarded more or less completely as spiritual rather than natural. A partial reduction of the three-fold division into a two-fold one took place in India as well. However, there, in general, the psyche has been considered in the realm of nature, and therefore subject to the laws of nature and amenable to scientific inquiry. Thus thoughts and feelings, and psychic phenomena, including those considered paranormal, are in the realm of *prakriti*, nature, that is to say, in the domain of materiality and causality. According to Indian thought, the so-called miracles, for example those mentioned in the Bible, are not supernatural or spiritual, even though they are unusual and extraordinary. Spirit is still beyond.

Evidence

We have already spoken about the somewhat obvious fact that our knowledge depends on the procedures, methods, and interests that we bring to knowing the cosmos. Neils Bohr was quite right in saying: "It is wrong to think that the task of physics is to find out how nature is. Physics concerns what we can say about nature."[3] Of course, even what we can say about nature depends on the mode of discourse a community of scientists accepts as the appropriate mode. In that universe of discourse only certain kinds of data are acceptable as evidence and certain other data are not acceptable. For example, the angels, so very real to Blake, are not acceptable scientific data, nor are Bach's fugues. In fact, no interior experience is a part of scientific data. Although one may speak in general of the scientific experience, it is necessary to distinguish between *experience* and *experiment*.[4] What we utilize in the sciences, and more particularly in the physical sciences on which our cosmology is based, are experiments and certainly not experiences. The words *experiment* and *experience* are derived from the Latin words *experimentum* and *experientia*, which in turn are both derived from *experiens*, the present participle of *experiri*, which means to try thoroughly, to risk, to go through; the clear implication is that this involves some personal participation and risk. We still use experience in this sense, but *experiment* has not been used in this sense for nearly three hundred years. *Experiment* is used these days as an intransitive verb, and no longer transitively as experience is. We can experience a flower, but we can only experiment *with* or *on* it. The scientific evidence about something is not gathered by exeriencing it but by experimenting upon it.

The knowledge thus produced is not a knowing-by-participation, but a knowing-by-distancing. It is not an I-thou knowing but an I-it one. Thus we see that scientific knowledge is indeed objective; but it is not objective in the mystical sense in which the observing self is so completely emptied or naughted that the object reveals itself as it is, the thing in itself, in all its numinosity and particularity. Sages in all cultures have said that it is in this state of consciousness devoid of the self alone that an object is known both in its oneness with all there is and in its distinct uniqueness. An entity – be it a tree, or a person or a culture or the whole cosmos – is then understood both in its interiority and its externality, including its generality and specificity. Scientific objectivity comes from another route, even etymologically, when we throw ourselves over and against something, as is understood in our word *objection*. One mode is that of love, the other of combat. Mystics, as everyone knows, are constantly speaking about love. We are told that God is love, as in the New Testament, or that love is what suppports the whole cosmos, as in Dante's *Divine Comedy*, or, as in the *Rig Veda*, that love was the first creation and absolutely everything else came from it. But by our scientific methods we wish to conquer nature as if she were an adversary. In fact, scientists almost never refer to nature as she; she is always called it.

Naturally, what is dead or was never alive can hardly have intentions, purposes, reasons, or feelings. In short, it can have no interiority. Evidence that involves this sense of interiority, that is based on an I-thou relationship, is out of the scientific arena altogether.

What is at issue here is a different sort of knowing. The important thing is not to see different things, but to see differently; not changed or expanded contents of the same consciousness, but a different quality of consciousness. Just as one can be in an I-thou relationship even with a cat or a tree, as Martin Buber used to say, one can also bring the I-it attitude to human beings, or even to God, if we seek only to use them as objects. Such, for example, was the attitude of Newton, perhaps the greatest of all scientists; as one of his biographers, Frank Manuel, has remarked, "For Newton, persons were usually objects, not subjects." The suggestion is not that scientists have any monopoly on the I-it attitude or that they are, as a class, devoid of the I-thou intercourse, but that in science, as distinct from some other possible activities of scientists, the I-thou attitude and any observations based on the inclusion of interiority of the object are automatically excluded from the body of scientific evidence.

In the last four centuries, there has been a virtual explosion in the number of scientific instruments that have extended our ability to observe the very small and the very far away and to measure extremely small amounts of time. In this immense quantitative expansion of the field of our observation, it cannot be said that we now see the cosmos with different eyes. There has been an extension of our eyes but not their cleansing, as Blake or Goethe would have understood it. There is nothing in the nature of science itself which might make one invoke, with St Francis, "Brother Sun; Sister Moon."

Any one of us can, of course, be deeply moved by a sense of our oneness with the cosmos, scientists as well as non-scientists. Furthermore, one can be struck by the wonder, the mystery, and the design of the cosmos as much today as in the days of Newton or Archimedes or Pythagoras, although unfortunately most of us are all too rarely. These feelings and perceptions lie in dimensions different from the ones in which our scientific observations are extended. No amount of quantitative expansion of data and theories can lead to the dimension of significance, any more than an endless extension of time can lead to eternity.

Design

It is hard to imagine a scientist who does not see order in the universe, a harmony of the various forces that permit the continued existence of the world, and a pattern involving regularity of phenomena and a generality of laws. The more we know about the universe, the more elegantly and wonderfully well ordered it appears. Most scientists share with Einstein a "deep conviction of the rationality of the universe," and his feeling that no genuine scientist could really work without a profound "faith in the possibility that the regulations valid for the world of existence are rational, that is comprehensible to reason." Einstein himself called this a "cosmic religious feeling," which he regarded as the "strongest and noblest motive for scientific research." Even though other scientists may be shy or embarrassed by the word *religious*, they are by no means strangers to the feeling that Einstein is describing.

What puts scientists on guard is not the idea or the feeling of design in the universe, but a suspicion that lurking behind the slightest concession in using the word is a theologian who will jump with glee and immediately saddle them with the notion of a Designer and all that goes with it. It is not the design that the scientists are uneasy about, but the designs that they smell hiding behind the slightest admission of it! It is no use telling them that the theologians have been on the defensive now for nearly three hundred years and are so eager to gain any approval from their scientific colleagues that they get a little over-enthusiastic if they sniff any possibility of truce. All of science is a celebration of pattern, regularity, lawfulness, harmony, order, beauty; in other words, all the marks of design. What it does not have much to do with is the Designer, who is over and above the design, occasionally interfering in the universe in contravention of natural laws. Already in the seventeenth centuy, Leibniz was able to remind Newton that his God was like a retired engineer: having created perfect laws and having set the universe initially in motion, He was no longer needed, and could be on a permanent sabbatical. The very perfection of scientific laws and their comprehensibility make the continued presence of this sort of God less necessary.

To have to infer the Designer from the design is largely a particular type of theological and linguistic habit. It is based on a notion of design that is more technological in character than scientific or artistic. In art, there is always present a definite element of play, improvisation, and surprise. No creative work is like painting by numbers; the artist does not know beforehand what the finished product will be like. And any scientist who already knows what he is going to find at the end of his work does not need a research grant, for he hardly needs to carry out the research. I am not discounting the intuitive conviction that a scientist can have about a particular idea or a method, so that he knows prior to engaging in a detailed calculation or an experiment what the outcome must be. But every good scientist, even an Einstein or a Newton, has many intuitive convictions that just do not lead any-where. In the actual working out of the ideas, and their encounter with what is, is the real delight, excitement, and even terror of creativity. Without them, scientific and artistic activity would be very dull. And any God who might create the universe without delight, without playfulness, without wonder, and without freedom and fresh possi-bilities would be a very dull God indeed. He would be a God of grim specialists, but not of the dilettantes, those who delight in what they do and study. Such a God could be a good technician carrying out a technical design, or a good bureaucrat keeping everyone in his place, or a thorough accountant keeping track of everyone's actions for later dispensation of necessary judgments; he might even make a good president of a large corporation, like a modern university. But he certainly would not make a good scientist, artist, or mystic. Such a God could not be the God of love or of wisdom, and it would be very difficult to take delight in Him.

Etymologically, *design* is also related to *sign from*. Sign from whom? Historically, in Christian theology, with rare exceptions, the signs are always from a personal God. However, there are profound and fundamental incompatibilities between scientific knowledge and the idea of a personal God, in spite of the fact that many very great scientists, for example Newton, were deeply committed to a personal God. Here is a brief excerpt from a manuscript of Newton, now in the Jewish National and University Library (Yehuda MS. 15.3, fol. 46^r):

We must believe that there is *one God* or supreme Monarch that we may fear and obey him and keep his laws and give him honour and glory. We must believe that he is the father of whom are all things, and that he loves his people as his children that they may mutually love him and obey him as their father. We must believe that he is Lord of all things with an irresistible and boundless power and dominion that we may not hope to escape if we rebell and set up other Gods or transgress the laws of his monarchy, and that we may expect great rewards if we do his will ... [T]o us there is but one God the father of whom are all things and we in him and one Lord Jesus Christ by whom are all things and we by him: that is, but one God and one Lord in our worship.

However, since Newton's time, and at least partly owing to the very science he took a major hand in creating, scientists are much less comfortable about accepting such a faith in a personal God, and certainly in expressing it. There is a feeling of a fundamental incompatibility between science and such a faith. Most scientists these days are likely to agree with Einstein in his description of what he called his religious feeling as one of

rapturous amazement at the harmony of natural law, which reveals an intelligence of such superiority that, compared with it, all the systematic thinking and acting of human beings is an utterly insignificant reflection ... The most beautiful thing we can experience is the mysterious. It is the source of all true art and science ... To know that what is impenetrable to us really exists, manifesting itself as the highest wisdom and the most radiant beauty which our dull faculties can comprehend only in their most primitive forms – this knowledge, this feeling, is at the centre of true religiousness. In this sense, and in this sense only, I belong in the ranks of devoutly religious men. (*Ideas and Opinions*, 1954)

Many people who knew Einstein personally insisted that he was the most religious person they had ever met. But he was not religious in any churchly or denominational manner. As he said, many times and in many ways, "My religion consists of a humble admiration of the illimitable superior spirit who reveals himself in the slight details we are able to perceive with our frail and feeble minds. That deeply emotional conviction of the presence of a superior reasoning power which is revealed in the incomprehensible universe forms my idea of God." Here we see a very good illustration of the fact that being struck by the beauty, harmony, order, and design in the universe does not necessarily mean accepting a personal or a sectarian God. It is worth quoting Einstein at some length on this point, from a remarkable address at a symposium in 1941:

The main source of the present-day conflicts between the spheres of religion and of science lies in this concept of a personal God. It is the aim of science to establish general rules which determine the reciprocal connection of objects and events in time and space. For these rules, or laws of nature, absolutely general validity is required – not proven. It is mainly a program, and the faith in the possibility of its accomplishment in principle is only founded on partial successes ... The more a man is imbued with the ordered regularity of all events the firmer becomes his conviction that there is no room left by the side of this ordered regularity for causes of a different nature ... To be sure, the doctrine of a personal God interfering with natural events could never be *refuted*, in the real sense by science, for this doctrine can always take refuge in those domains in which scientific knowledge has not yet been able to set foot.

But I am persuaded that such behaviour on the part of the representatives of religion would not only be unworthy but also fatal. For a doctrine which is able to maintain itself not in clear light but only in the dark, will of necessity lose its effect on mankind, with incalculable harm to human progress. In their struggle for the ethical good, teachers of religion must have the stature to give up the doctrine of a personal God, that is give up that source of fear and hope which in the past placed such vast power in the hands of priests. In their labors they will have to avail themselves of those forces which are capable of cultivating the Good, the True and the Beautiful in humanity itself. This is, to be sure, a more difficult but an incomparably more worthy task. ("Science and Religion," in Einstein's *Out of My Later Years*, 1950)

In my judgment, which in this regard is different from Einstein's, the major cause of the incompatibility beween science and theology or churchly religion, which should certainly not be confused with spirituality, is not so much the concept of personal God *per se* as the restricted view of knowledge that prevails in scientific circles, as remarked earlier, and the limited notion of the Spirit or Divinity that the theologians have. To have understood rightly that Divinity is at least at the level of the human person, does not mean that it is only personal. The personalist aspects of Being, such as intelligence, intention, will, purpose, and love, which are all marks of interiority, do not have to lead to a concept of a personal God made in the external image of man, with definite form and being separated from the others. Uniqueness of any level of being, seen separated from the oneness of all Being, leads to a limitation of vision, to partiality and exclusivism. As the scriptures tell us, man is made in the image of God, which I take to mean that man is potentially able, in the deepest part of himself, to be one with the Divine. This is what the sages have always said, everywhere, whether the expression is *aham brahmasmi* or "My Father and I are one." However, if we forget the summons for an inward expansion to God, we are bound to reduce God in an outward contraction to man.

Concluding Remarks

I have argued that there is more to the universe and to knowledge, and the corresponding evidence, than is encountered in physical cosmology; that there are dimensions of the existence and development of being other than in time; and that one can be very spiritual with a personal God or without one. These are practically truisms. In any case, my observations have nothing to do with being Eastern or Western. Of course, everyone is conditioned by one's cultural background. However, the more deeply one delves into oneself, the more one discovers one's common humanity with others, and one's commonness with all there is, without thereby losing one's uniqueness. In this necessary realization of our oneness as well as uniqueness, we may, each one of us, have to travel paths we do not ordinarily travel, in lands we do not usually inhabit, and experience modes of being not habitually ours.

Different modalities and levels of being, and the corresponding levels of thought and feeling, exist in every human being and even more so in every culture. Some contingent, historical factors can overwhelm or underscore a particular modality at any given time. The tremendous impact of science and technology in the West in the last two centuries has made some modes of being now appear to be non-Western. Yet we are now in a particularly exciting situation of a global neighbourhood demanding a larger vision of ourselves. A special kind of insensitivity is now required for us to remain culturally parochial, refusing to become heirs of the great wisdom of mankind: as much of Plato as of the Buddha, of Einstein as well as Patanjali, of Spinoza no less than that of Confucius.

A major conceptual revolution was created in the Western world when the works of Aristotle were discovered by the Latin West through the Arabic philosophers in the eleventh and twelfth centuries. The revolution went on for several centuries, leaving no area of thought and culture untouched. It appeared for a time that the major synthesis brought about by Thomas Aquinas between Aristotle and Christian thought was a culmination of this revolution. But no: it rolled on until and including the major scientific revolution of the sixteenth and seventeenth centuries that was finally brought to a close by Newton. Since the end of the nineteenth century, we have been in the middle of another very major encounter of different cultures and different streams of thought, of the West with the East. There is, moreover, an important aspect of the contemporary situation, since the Second World War: for the first time in history major cultures are juxtaposed as neighbours without being in the position of either the victor or the vanquished. Who knows where the resulting cultural revolution will end?

One thing, however, is certain: a consequence of this revolution is bound to be a recognition, in addition to the experimental science of nature that has been a particular achievement of the modern West, of an experiential science of the Spirit freed from all sectarian theology. This science of the Spirit is not the same thing as an extension of our present science to include occult phenomena and extrasensory percep-tions. Also, one should not let oneself be seduced by superficial parallels between certain expressions and paradoxes of contemporary science and ancient oriental thought. It is true that here and there are beginning to appear, in the long column of Western appelations in the honour rolls of science, names like Chandrasekhara Venkata Raman, Tsung Dao Lee, Hideki Yukawa, Abdus Salam, and Chen Ning Yang. In his day Kepler was convinced that the Sun was the Father, the circumference of the solar system the Son, and the intervening space the Holy Ghost. A latter-day scientist, brought up on different symbols and metaphors, might see in the patterns appearing in the cloud chamber the dance of Shiva, or be moved to find in the complementarity appearing in the quantum phenomena *yin* and *yang* encircled together, or discover the resolution of the various paradoxes of contemporary physics in the ineffable Tao. These parallels or interpretations are as true or false now as they were then. They add nothing, either to true science or to true spirituality.

There is a deep-seated need in human beings to seek an integration of all their faculties, and a unity of their knowledge and feeling. We are fragmented and thirst for wholeness. This thirst, however, cannot be quenched by mere mental conclusions and arguments about the parallels between physics and Buddhism or about the existence and nature of the design in the cosmos. What we need is a radically transformed attitude – in the deepest sense, including the posture of the body, as is happily conveyed by the corresponding French word--which would permit us to receive true wisdom and intelligence from above ourselves, and to use our science and technology with compassion and love. Without this attitude we cannot reconcile Blake and Newton, and their future heirs. And the lament will continue:

> O Divine Spirit sustain me on thy wings!
> That I may awake Albion from his long and cold repose.
> For Bacon and Newton sheathd in dismal steel, their terrors hang
> Like iron scourges over Albion, Reasonings like vast Serpents
> Infold around my limbs, bruising my minute articulations ...
> In heavy wreathes folds over every Nation; cruel Works
> Of many Wheels I view, wheel without wheel, with cogs tyrannic
> Moving by compulsion each other: not as those in Eden: which,
> Wheel within Wheel, in freedom revolve in harmony and peace.
>
> (William Blake, *Jerusalem* 15: 9-20)

The tension between the two major contributing streams to the Western mentality, the Greek and the Hebraic, with their respective emphases on the cosmological and the theological perspectives, is very old. Whitehead once remarked that this tension may have been the main source of the creative dynamism of the Western culture for centuries. However, now there is almost a complete separation between these two perspectives. As I said earlier, scientists are deeply committed to the cosmological perspective to the exclusion of the person. On the other hand, neither theology nor philosophy has been an experiential science for many a century. If we were to take the corresponding experience seriously, as with great mystics and spiritual masters, one thing would become immediately clear: there is not much meaning to consciousness or intelligence, and thus to God, without the accompanying attributes of action, love, and delight. These are not so much attributes of the Spirit, added from the outside and without which the Spirit could exist, as they are the means by which we recognize the presence of the Spirit or Consciousness or Intelligence. Thus God is not only omniscient, but also omnipotent, omniamorosus and omnidilettante! This is precisely what make the Spirit omnidelectabilis so that human beings are constantly drawn to Her and in love with Her. Occasionally, they write poems of ecstasy for the Spirit, as did Alexander Skryabin, a Russian composer and poet of the early twentieth century:

> The Spirit playing,
> The Spirit longing,
> The Spirit with fancy creating all,
> Surrenders himself to the bliss of love ...
> Amid the flowers of His creation, He lingers in a kiss ...
> Blinded by their beauty He rushes, He frolics, He dances,
> He whirls ...
> He is all rapture, all bliss in this play
> Free, divine, in this love struggle
> In the marvellous grandeur of sheer aimlessness,
> And in the union of counter-aspirations
> In consciousness alone, in love alone,
> The Spirit learns the nature of His divine being ...
>
> *(Poem of Ecstasy)*

Thus the design, and the intelligence in it, behind it, and above it, turns into a dance. Some dancers come and go, join the dance, or stop to watch it; but the dance goes on eternally, in the beginning as now, in love and delight. In moments of wholeness, of deep feeling and clarity of awareness, each dancer is unique in himself and one with the cosmos, and he knows that In the Beginning Is the Dance of Love.

NOTES

1 In this connection, see A.H. Armstrong and R. Ravindra, "The Dimensions of the Self: *Buddhi* in the *Bhagavad Gita* and *psyche* in Plotinus," *Religious Studies* 15 (1979): 327–42.

2 We see a good example of this identification in French: the last words of Jesus Christ before his crucifixion, in the Gospel according to St Luke 23:46, are translated into French as "Père, je remets mon esprit entre tes mains." The word *esprit* now means in English both *spirit* and *mind*. It was pointed out to me by the French physicist Jean Charron that Descartes in *Meditations* VI seems to make some room for an entity higher than both the body and the soul. Perhaps; but the Cartesians?

3 Quoted in Ruth Moore, *Neils Bohr* (New York: Knopf, 1966), 406.

4 In this connection see R. Ravindra, "Experiment and Experience: A Critique of Modern Scientific Knowing," *Dalhousie Review* 55 (1975–6): 655–74.

TERENCE PENELHUM

Science, Design, and Ambiguity: Concluding Reflections

It is not my purpose here to recapitulate the papers in this volume, but to offer one view of the way in which they advance the perennial debate of which they form a part. To do so, I shall, first, try to say where the debate stood up to our time – that is, after the advent of Darwinism, but before the spectacular developments described here. I shall then estimate the impact of current scientific findings on the status of this debate, and conclude with some observations on the attitudes of my humanistic colleagues, and a tentative judgment of my own.

I

However deep the disagreement about the claims of the Judaeo-Christian religious tradition, the belief in a creator deity which is at the centre of that tradition has completely determined the thinking of each of us in a fundamental respect.

In his introduction, Dr Batten connects the difficulties of the traditional Argument from Design with the difficulties of the belief in final causes. Of course, they *are* connected. But the nature of their connection has to be recognized.

The belief in final causes comes from Aristotle. To say there are final causes (or natural teleology) is to say that natural processes fulfil certain purposes, or achieve certain ends; it is also to say, at least for Aristotle, that this fact is irreducible, in that one's understanding of such processes is incomplete if those purposes are not known, and that it is as basic scientifically that these ends are achieved, as that the processes which achieve them have causal antecedents. But Aristotle did not maintain that the presence of such final causes shows that there must be any conscious agent, or agents, who entertain or impose those purposes. (The God in whom Aristotle believes is not a person or an agent, and is quite unaware of the purposes nature satisfies; he merely serves as an object of unconscious imitation for terrestrial beings.) Similarly, when the Stoics used the Argument from Design, they took themselves to be establishing the presence of an immanent cosmic Reason, not a personal deity. It is only when the doctrine of final causes is combined with a theism that comes from other sources, that we find it assumed that the presence of final causation in nature requires the postulation of a distinct conscious mind to entertain the purposes nature is thought to serve and direct natural processes to their fulfilment. I say "assumed," because this does function as an *assumption* in virtually all later versions of the Argument from Design. (Aquinas states it explicitly in his Fifth Way: "Now things which lack awareness do not tend towards a goal unless directed by something with awareness and intelligence, like an arrow by an archer."[1] But he saw no need to justify it, and his harshest critics have never suggested he should have.)

This is a radical *theistic turn* in cosmological thought. It is accepted both by those who believe in God, and by those who do not. The most conspicuous result of it has been that whenever scientific discourse, in biology or elsewhere, appears to make use of purposive language (in descriptions of modes of adaptation, for example), those who speak this way feel obliged to insist that such language is either metaphorical, or is a convenient shorthand for statements that would contain no purposive vocabulary. For it is feared that the acceptance of final causation would entail the conclusion that there is a God who imposes it. Since such a conclusion is not a scientific one, the language of science is judged to be necessarily non-teological. If it were not, the very autonomy of science would be felt to be threatened, and it has been an imperative of our culture since Galileo that that is unaccept-able. There has been, paradoxically, a close connection in our civiliza-tion between the dominance of theism and the death of natural teleology.

I will express this point in another way, using the language of this volume, *fine tuning* instead of final causation. Aristotelian science insisted that every substance, because of the sort of being it was, was fine tuned intrinsically, so that its behaviour and development would issue in certain results. Since the end of medieval times, it has been assumed that if any facts about the universe or its inhabitants are fine tuned, they must be fine tuned from outside. This assumption has nourished theism in early modern times, and has assisted the transition from medieval to modern science. But that transition has not been without cost: the judgment that fine tuning is present, since it entails a theological conclusion, represents a judgment that some phenomena can be understood only by passing beyond what they show to be scientific limits. The acknowledgment of fine tuning is a concession of scientific defeat – a gap.

The traditional, eighteenth-century Argument from Design most commonly took the form of saying that there are clear signs that our world has been fine tuned, even though, as Dr Batten rightly insists, it was also said to be a mechanical world. The signs of such fine tuning, and therefore of God's planning, are the phenomena of adaptation. William Paley, for example, in his *Natural Theology*,[2] mounted an enormously influential version of the argument by detailing a large number of examples of such adaptation. These were judged to show that the physiology and instinctive behaviour of organisms fulfil purposes, because they serve the *good* of those organisms. If this is true, there must be an external directing intelligence who has endowed them with this physiology and these instincts, since they cannot have done so themselves.

Paley's argument, as Dr Haynes has reminded us, was just one in a long line of repeated versions of the same intellectual move. It is important to remember that this move was thought to be overwhelmingly obvious by all but a very few people, so that even those who did most to discredit it, namely Hume and Kant, felt obliged to do something to explain why it seemed so obvious to everyone but them. It is noteworthy, for example, that Paley wrote his *Natural Theology* after Hume's death, and completely ignored all Hume's criticisms of the Argument from Design, even though he took pains to refute Hume's views on other matters. It was thought, in fact, to be clearly irrational not to see the adaptations around us as the manifestations of the creator's planning.

Hume's criticisms are now generally regarded by philosophers as sound. Among them is a speculative anticipation of Darwin, which appears in Part VIII of the *Dialogues Concerning Natural Religion*.[3] Hume suggests that adaptation might begin by accident, and then be self-perpetuating. My own favourite criticism of the Argument from

Design is to be found not in Hume, but in Voltaire's *Candide*, where Dr Pangloss extols the beneficence of God's design in making our noses exactly the right shape to support our spectacles. But it was Darwinism that damaged the Argument from Design most radically. Darwin offered a non-teleological explanation of the adaptations we observe, an explanation which is inevitably more economical, and is supported by a vast range of independent evidence. It is not a mere speculative possibility like Hume's, but is confirmed by an overwhelming variety of relevant data. This same evidence also demonstrates beyond reasonable doubt that there has been an immensely long history of maladaptation that has yielded suffering and extinction on a vast scale. So the facile optimism that underlies the estimate of natural history in the Argument from Design no longer has any excuse, and its conclusion, if not proved false, is at least shown to be unnecessary in accounting for the facts said to require it, as well as shown to have evidence against it.

In the wake of this change, the most a rational theist seems able to do is to retreat from the enterprise of natural theology and engage instead in the more modest enterprise of defensive apologetics. The task here is to argue that the world Darwin has shown we inherit is one whose character can be *reconciled* with the belief in creation by a loving and omnipotent God. This is a very different undertaking: instead of contending that the evidence places the existence of a creator beyond reasonable doubt, the task is merely that of showing that believing in a creator is not *un*reasonable; that someone who believes in God does not have to disregard the evidence.

I think many religious believers today would settle for this. I also think that defensive apologetics, thus understood, has been largely successful. But where does it leave us? It leaves us in a position where the believer and the sceptic each holds a view that the facts do not refute, but also do not require on pain of irrationality--even though only one of their positions can be the truth.

On approaching this discussion, my view was, roughly, that this state of equipoise is our actual intellectual situation: that neither theism nor atheism is irrational in relation to the evidence, and that neither is required by it. I shall call this estimate of our situation the view that our world is *religiously ambiguous*. I think this is where we came in. The question is whether anyone has now shown reason to suppose that the ambiguity is resolved. Has anything been discovered which would tip the scale decisively in one direction or the other? Can any of these scientists or philosophers *disambiguate* my world?

II

In spite of the great variety and complexity of the scientific developments that have been described here, there is a striking similarity between the messages stated or, more commonly, implied in these expert accounts. Each paper deals, directly or indirectly, with the understanding of one remarkable fact: that the universe in which we live has evolved in such a manner that it has produced the conditions necessary for the development of intelligent life – specifically, of ourselves. As Dr McLaren puts it, the thrust of each contribution is anthropocentric. At first glance, this emphasis seems arrogant and self-serving. There is no good reason to suppose that a universe that had not evolved us could not have been designed: indeed, since so much of the universe we inhabit could not sustain us, some of it must have a point that excludes us even if it *is* designed. But we should probably not feel sheepish about this anthropocentrism, since if there is design and we are intended to be aware of it, it would be odd if our own place in it were not specially remarkable. The point was not lost on Pascal, who found it necessary to balance the shock of the infinity of empty space with the perception of the uniqueness of human awareness.[4]

But although each contribution has conceded that the evolution of human life is cosmically remarkable, the consistent message is that it is not recalcitrant to explanation in terms of existing, or imaginable, theories. Throughout the contributors have stressed the importance of particular conditions that had to be satisfied before intelligent life as we know it could have come about, and have then shown how, as far as we can now tell, Nature surprises us, as Peebles puts it, without bending the rules. Examples of such conditions included the balance of potential and kinetic energy (Peebles); the proton-antiproton abundance asymmetry (Unruh); the coagulation of the matter of the universe into stars and planets with appropriate lengths of life and atmospheres (Ovenden); the occurrence of the mutations necessary for the processes of selection to work upon (Haynes); the emergence of intelligence and the use of language (Kimura). The presentations are rich in illustrations, both central and incidental, of the explanatory resources of contemporary scientific thought, even in areas where those resources are most obviously in process of development: the growing understanding of the interrelation of biological and geological evolution (Veizer); the impact of major extinctions on the course of organic development (McLaren); the recognition of the degree to which life, once present, contributes to its own stability and perpetuation (Lovelock); the extent to which human linguistic capacity has biological antecedents and counterparts (Kimura). In each case, some condition for the emergence of intelligent life is addressed because it seemed to require explanation: more precisely, because without such an explanation from within that contributor's discipline its presence might suggest external fine tuning. Each seemed then to say that the phenomenon that is so remarkable when first recognized is not, after all, so very surprising when we learn what the rules permit. So no external fine tuning appears to have been necessary. Perhaps molecular biology supplies the most striking recent examples of the dispensability of teleological hypotheses (see particularly Glickman and Doolittle), but in this respect it is merely the source of the most *striking* examples. From all the scientific disciplines represented here, the message has been the same: none of these conditions for the emergence of intelligent life opens what Meynell calls a gap for God.

I would go further: with the possible exception of Professor Doolittle, there is not even any trace of the "mal de l'immensité" of which Dr Audet writes so eloquently, because the infinite spaces of which Pascal wrote a quarter of a millenium ago are not silent for us any longer. A language has been developed to write about them, and it is a language which, though very different, as Dr Audet says, from the language of classical Greek ontology, is still resolutely non-teleological. While humanity, or at any rate scientific humanity, may not yet see its way clear to controlling these vast spaces, intellectually it has already domesticated them.

For many, such a verdict generates uneasiness. Puzzlement and surprise, even anxiety, are, after all, more the driving force of science than of theology. But although no theorist can set psychological limits to what will amaze us, the persistence of feelings of amazement does not legitimize every demand for special explanation. I can think of no better series of reminders of this important truth than Professor Hacking's demonstrations that there is, as he puts it, no *general* answer to the question of how surprising a coincidence is. To create the sort of gap Dr Meynell speaks of, one would have to produce some reason of principle for supposing that this or that fact that has surprised someone could not be subsumed under existing theories or models. He points out that there are teleological difficulties in the use of such supposed gaps, in that they suggest a less omni-competent designer than God is traditionally supposed to be. (He ought to be careful, though, in using an argument that was the stock-in-trade of the eighteenth-century deists, who employed it to deny the possibility of divine interventions in history.) But aside from that, gaps have an unpleasant habit of closing before you can ride through them, and in these papers seven or eight of them snap shut in a flash.

It seems to me that the effect of this impressive record of scientific achievement is to cast very grave doubt on the possibility of reviving the Argument from Design by stressing the presence in our universe of the necessary conditions for intelligent life. To say this is not to make it difficult for one who believes in God independently to discern his handiwork in life's development, or the signs of his intelligent contrivance in the very laws that make its development intelligible. Intelligibility does not have to prevent awe and wonder, any more than familiarity has to breed contempt. But there is a clear sense in which the intelligibility that science continues to promise renders the belief in design unnecessary. The religious ambiguity remains.

III

A schematic summation like the above is unjust, since it has to omit opinions which do not fit neatly into the pattern I have claimed to find.

1 Consider a theme emphasized particularly by Dr Haynes, one I address with anxiety, because I am the merest dunce about it. I refer to his insistence that there is a critical factor of chance in those mutations required by natural selection. This factor is one he ascribes to the sort of indeterminacy that is a feature of quantum phenomena, and is therefore not merely a mask for the current inability of investigators to make predictions. Professor Swinburne has suggested in discussion that this can be accommodated easily enough into a belief in design, but I am uncomfortable. At least it seems to me that we cannot think of design as absorbing chance, even though such a suggestion has tempted theologians.[5] The idea of planning to do something by chance looks incoherent: chance is not a method or procedure, and the outcome of that which is genuinely random is, for that very reason, not planned. Of course God, like any one of us, could *leave* an outcome to chance, but I find it hard to combine it with the claim that he has thereby *planned* that outcome. The best one could mean by saying that chance can be absorbed in design is that ultimately it is only apparent and not real, and I suppose that is what Professor Swinburne meant. If one comes to the phenomena with a prior conviction of God's existence, such a claim is no doubt possible; but when indeterminacy is recognized as scientifically ultimate, the judgment that it is somehow

mere appearance cannot, of course, be offered as having a scientific grounding. In my view, the pervasiveness of chance is a strong *prima facie* candidate for a fact that could remove religious ambiguity, and it is one which, if it removed it, would remove it negatively; but the theistic move I have allowed to be possible does look like a way of evading it at the cost of unverifiability.

2 Professor Swinburne contends that there is indeed a prominent fact that we have reasons of principle to think is beyond the explanatory resources of science. This is the fact of consciousness. He acknowledges we can readily offer reasons why beings with brains like ours have survived, and can invent reasonable speculations why the consciousness associated with the processes that occur in such brains gives their possessors evolutionary advantages, but insists there is a deeper residual puzzle: why do neural processes give rise to the conscious events they do, and vice versa? He maintains that this explanatory gap is one which can best be filled by invoking the personal explanation of the will of the deity. I am inclined to accept that a scientific explanation of these correlations is hard to envisage, let alone confirm, and I share Swinburne's view that the sophistries of eliminative materialism do nothing to dispel the stark conceptual disparity between descriptions of neural and conscious states. But what sort of gap is revealed here? Clearly it is indeed one that seems to block the creation of an overarching theory of high generality that can accommodate consciousness within it. Hence the non-theist must rest content with the emergence and persistence of consciousness as a brute fact whose conditions can be described, but not, in this deeper sense, be understood. I happily concede that reference to design provides an additional explanation not otherwise available, but I am unconvinced that this explanation is either compelling, or even, as Professor Swinburne would prefer to say, more clearly probable. For the theistic explanation is in terms of the will of an immaterial being of limitless power and (to the human mind) of infinite complexity, whose being must be taken as a sheer, ultimate, fact in the way in which the traditional materialist took that of matter to be a sheer, ultimate, fact. If the ultimate fact is taken to be immaterial, then the existence of matter needs explanation; to say that this explanation resides in God's power, even if this is true, is not to make the existence of matter theoretically intelligible. It is merely to say that God has the power to cause it, which does not tell us how this comes about – an enterprise which would in any case be religiously objectionable. Treating the emergence of consciousness from matter as ultimate in a similar way is indeed

puzzling and unfulfilling, but it is not clear that closing this gap in the way Swinburne suggests avoids a parallel puzzle.

3 Professor Meynell has emphasized a fact which, in his view, requires explanation, and is a more significant sign of design than any other. It is more significant because it remains, in his view, even if one concedes that there are no explanatory gaps within science into which theistic explanations can be inserted. It is the fact of the world's intelligibility itself. Why is it that there is such a remarkable "fit" between our cognitive demands and the world we investigate? As he says, Kant answered this question by telling us that the sensory input we receive from the world is filtered through a complex cognitive mechanism as it enters our consciousness; and he drew the conclusion that we can only know the world as it looks through this filter, not as it is. I have my doubts about the question itself. The idea that our minds and the world have to "fit" like a cap and bottle, before we can understand the world, seems to be a relic of discredited correspondence theories of truth. If we express the problem without using this image, it seems to reduce to the question of how it comes about that we are able to understand our environment. This question, I am afraid, still can get a Darwinian answer: namely, that survival is due to adaptation, which in turn is the result of mutations. Most of these mutations have led to behaviour patterns that preserve those who manifest them without endowing them with any cognisance of their environment. In our case, what has kept us going is our chance capacity to acquire some beliefs and to act on them. True, this is remarkable, but so are all the other adaptive mutations when you think about them.[6]

4 I turn finally to a suggestion that arises, if I understand him correctly, in Professor Ravindra's paper, and is incorporated in its title. It is a theological suggestion: that we would do well to absorb something of the Hindu conception of divine creative activity, symbolized by the dance of Siva. This is a conception of creation as play (or *lila*). There are Judaeo-Christian precedents for this suggestion, and it is one that I would certainly welcome for independent reasons. But I am afraid it does not help to resolve the question we are addressing. It implies, what I think is healthy, that the sheer richness and variety of life-forms is itself an expression of the creator's act, so that we are in no position to suggest that he would be unfulfilled if he had created others but not us. It also incorporates a view of the creator's love that emphasizes delight in the creatures, rather than need of them. These are valuable lessons for Western theism to learn, or perhaps to relearn. There is no doubt, either, that chance or randomness seems to fit in

intuitively with this image of creation more readily than with the image of a premeditative creator who sets in motion a long process that culminates inexorably in the emergence of the Royal Society. But unfortunately, these merits are the wrong ones for us here. The very aptness the image has for reconciling creation with chance makes it unsuitable for reconciling chance with design. It gives us an opportunity of thinking up a theology in which our world is the by-product of its creator's high spirits, not of his intellect. The dance of Siva, if you like, is not choreographed.

IV

If what I said in my first section is true, the message I have claimed to find in our scientific presentations in my second section is not surprising, but is the only one to be expected from disciplines that continue to be so spectacularly successful. If my brusque and compressed judgments in the third section are correct, the continued ambiguity with which that message leaves us has not been removed. But I do not think that this deflationary opinion tells us more than half the truth.

The religious ambiguity in which I think we find ourselves is one that carries heavy costs with it. Everyone has to resolve it individually. Either you decide you live in a world governed by God, or you decide you do not. One true thing in theological existentialism is Pascal's assertion that not deciding is equivalent to deciding negatively.[7] If you decide positively, the intellectual costs are substantial: you have all the painful mysteries that do not fit easily with belief in God, such as the enormous incidence of death and suffering and extinction in the world's biological history. If you decide negatively, the costs are substantial too: you have the relative austerity and lack of explanatory richness and depth in a world where science confronts you with sheer chance. We would all like to know which of these worlds is ours, so that we know which price we ought to be paying.

Although I am far less optimistic than Professors Meynell and Swinburne about the potential success of natural theology, I must agree with both that rational believers must reject all forms of fideism – the perverse theological doctrine that tries to make a virtue out of natural theology's failures, and tell us it is a good thing for faith and reason to have nothing to do with one another. I think, therefore, that those who have looked for gaps in the scientific picture have been right to search for compelling reasons to break our intellectual deadlock, even if they have been unsuccessful in finding them.

The way they have tried is significant. They have approached key turns in our evolutionary development in a way similar to that in which historians approach puzzles about our political or social past – puzzles about the cause of key events that have had momentous consequences. To notice this is to recognize a special feature of post-Darwinian science, and of contemporary cosmology. The old mechanical model of the universe that was used in the classical Argument from Design implied that we live in a world whose basic features have always been the same, except for those changes that we humans have brought about in it. It was only from revelation that we learned that it began in time. So history is just *our* history. What we now see, and have learned from science, is that the world, also, has a history: a datable beginning, and a subsequent development through critical and irreversible stages. I do not wish to argue that this awareness does anything, by itself, to undermine the classical assumption of universal natural law. But a world with a history is a world which fits, even though it indeed does not require, the eschatological orientation of the major western religious traditions. The trouble with the classical Argument was that the more its proponents convinced their hearers of the world's orderliness and predictability, the more they also convinced them of the irrelevance of God, if he exists, to any particular sequence of events. The God of the classical Argument is the god of deism, not of Judaism or Christianity or Islam – a fact that the fideists have seen, but have misused. What those who try to insert God's agency into gaps in the world's history are trying to do is to duplicate for the history of the cosmos something the western religions have traditionally claimed to be true only of certain selected events in *human* history. They have tried to suggest that they are due to divine interventions, and that without those interventions we would have had the best possible reasons for expecting quite different events to have happened. They have been looking for cosmological miracles. I do not say that there can have been none; but when one considers the candidates, one returns to ambiguity in the way we have seen. These events can be viewed as striking manifestations of the providence of God if you already believe in the providence of God, but they do not make it unreasonable *not* to believe in the providence of God.

If there have been any miracles in human history, remote or recent, they have been different. If you look at the miracles believers say have happened, they have been events that do make it unreasonable to deny that God has been at work. (I am not saying that if they happened, no one *would* make that denial.) They are, or were, special events that teach certain things about God to those who are witnesses to them. Each has a special pedagogical setting, which uniquely fits the needs of those before whom it (allegedly) takes place. If they reject its message, then they are being perversely unreasonable. A part of that pedagogical setting is that if the miracle happens, any rational being of normal cultural development sees that the laws of nature have been violated. We do not need an Einstein to tell us whether or not someone's walking on water or rising from the dead violates a natural law. Such things, if they happen, are unambiguous events. (So unbelievers have no choice but to deny that they occur.)

So if we are hoping to find facts that place God's presence beyond reasonable doubt, rather than merely something which there may be a slight edge in favour of, or against (so that religious ambiguity continues), it seems to me unlikely, on the evidence in this volume, that we could find them in the origin or evolution of the cosmos. Hence I am pessimistic about the prospects of standard forms of natural theology that try to establish the being of God in Stage One and proceed to specific religious doctrines about God in Stage Two. I think the whole weight would have to be carried by theologically loaded miracles. There are many well-known arguments designed to show it is not rational to believe in them, but these are not our present concern.

Meanwhile, the cautiousness of my judgment could well be shown to be the grossest error tomorrow. The exciting scientific era in which we live is one in which openness is the only rational attitude, so even caution should be qualified.

NOTES

1 St Thomas Aquinas, *Summa Theologiae*, Part 1, Question 1, Article 3.

2 William Paley, *Natural Theology*, in *The Works of William Paley*, ed. Edmund Paley (new ed., London: William Smith, 1838), Vol. 1.

3 Hume, *Dialogues Concerning Natural Religion*, ed. Norman Kemp Smith (2nd ed., Indianapolis Bobbs-Merrill, 1947), 182–7.

4 Pascal, *Pensées*, Fragments 200 and 201 (Lafuma numbering).

5 See, for example, John Hick, *Biology and the Soul*, Eddington Memorial Lecture (Cambridge: Cambridge University Press, 1972).

6 I cannot pursue these difficulties here, but perhaps I may be allowed a short footnote. Professor Meynell's argument, developed more fully in his book, *The Intelligible Universe* (London: Macmillan, 1982), should be

contrasted with the position I think is implied in Dr Ovenden's conclusion. Epistemological scepticism has always been based on the recognition
that our sensory and intellectual apparatus inevitably mediates our
apprehensions of the universe about us, and the sceptic suggests that, in
consequence, we are not entitled to suppose that the apprehensions so
mediated coincide with the actual structures of the universe. Short of the
mind-twisting suggestion that we could aspire to some apprehension
that is not mediated either by the senses or the intellect (the claim of the
mystics), one can only aspire to apprehend the world as it really is after
absorbing the sceptic's doubts by appealing, in Cartesian fashion, to God.
One reason I have for not being inclined to do so is the fact that I do not
see that a world in which our science is fruitful and productive even
without such a guarantee is self-evidently *not* designed.

7 Pascal, *Pensées*, Fragment 418.

Contributors

JEAN-PAUL AUDET, S.R.C.
Professeur titulaire, Département de philosophie
Université de Montréal, Montréal, Québec

ALAN H. BATTEN, F.R.S.C.
Senior Research Officer, Herzberg Institute of Astrophysics
Dominion Astrophysical Observatory, Vancouver, British Columbia

W. FORD DOOLITTLE
Fellow, Canadian Institute for Advanced Research
Professor of Biochemistry
Dalhousie University, Halifax, Nova Scotia

BARRY W. GLICKMAN
Associate, Canadian Institute for Advanced Research
Professor of Biology
York University, Toronto, Ontario

IAN M. HACKING, F.R.S.C.
Professor, Institute for the History and Philosophy of Science and
Technology
University of Toronto, Toronto, Ontario

ROBERT H. HAYNES, F.R.S.C.
Council Member, Canadian Institute for Advanced Research
Distinguished Research Professor of Biology
York University, Toronto, Ontario

DOREEN KIMURA
Professor of Psychology
University of Western Ontario, London, Ontario

JAMES E. LOVELOCK, F.R.S.
Independent Scientist
Coombe Mill, St Giles-on-the-Heath, Launceston, Cornwall, England

DIGBY J. McLAREN, F.R.S., F.R.S.C.
President, Royal Society of Canada 1987–90
Visiting Professor of Geology
University of Ottawa, Ottawa, Ontario

HUGO MEYNELL
Professor of Religious Studies
University of Calgary, Calgary, Alberta

MICHAEL W. OVENDEN, F.R.S.E., F.R.S.C.
Late Professor of Geophysics and Astronomy
University of British Columbia, Vancouver, British Columbia

P.J.E. PEEBLES, F.R.S., F.R.S.C.
Associate, Canadian Institute for Advanced Research
Albert Einstein Professor of Physics
Princeton University, Princeton, New Jersey, U.S.A.

TERENCE PENELHUM, F.R.S.C.
Council Member, Canadian Institute for Advanced Research
Professor of Religious Studies
University of Calgary, Calgary, Alberta

RAVI RAVINDRA
Professor of Comparative Religion and of Physics
Dalhousie University, Halifax, Nova Scotia

RICHARD G. SWINBURNE
Nolloth Professor of Philosophy of Christian Religion
University of Oxford, Oxford, England

WILLIAM G. UNRUH, F.R.S.C.
Lac Minerals Fellow, Canadian Institute for Advanced Research
Professor of Physics
University of British Columbia, Vancouver, British Columbia

JÁN VEIZER, F.R.S.C.
Professor of Geology
University of Ottawa, Ottawa, Ontario